Public Consultation and Community Involvement in Planning

Public Consultation and Community Involvement in Planning is the definitive introduction to public consultation for developers, students and planners. The past decade has seen a complete transformation in consultation and community relations in the UK, from increased requirements to consult, to the introduction of neighbourhood planning and a revolution in online communication. *Public Consultation and Community Involvement in Planning* takes readers through consultation from the basics right through to emerging trends to demonstrate how a successful consultation process can benefit both the developers and the local community.

The book begins with a definition of consultation and community involvement and an explanation of their role within the development process, before going on to clarify the legal, ethical, practical and ideological concerns to be addressed by the consultation process. Consultation strategy is explored step by step, and social media and online consultation is explored in detail. This is the first comprehensive guide to modern public consultation within the UK development sector and will be essential reading for developers, students and planners.

Penny Norton is the founder and director of PR consultancy PNPR (www.pnprlimited.co.uk) and of ConsultOnline (www.consultonlineweb sites.com), a comprehensive online service designed specifically for use in consultations on planning applications. Penny has substantial experience in public relations and public affairs within property, construction and regeneration and has written extensively on the subjects of consultation and community relations.

Martin Hughes is a former Labour councillor. He founded specialist community involvement consultancy Polity Communications in 2005 and since then has dealt with over 100 local authorities, gaining a wide breadth of experience in development and regeneration projects.

'This book is a very comprehensive overview of the historical background, the obstacles and the democratic legitimacy challenges that come with consultation and community involvement in the changing world of planning, community relations and corporate social responsibility. It offers a tried and tested approach to citizen engagement strategies, informed by a strong theoretical grounding and years of experience on the ground, and provides excellent insights into best practice, demonstrated through a range of projects.'
—Adrian Penfold OBE, Head of Planning, British Land

'Perhaps the most remarkable change in planning since the 1960s has been the growing appetite for ordinary people to want to get involved in the planning process. Yet for too long developers have just played lip service to this valid interest. Having worked with Penny over the years I have greatly admired her approach and here it is described both theoretically and through wide-ranging examples. It is the antithesis of the "dark art" of spin so often associated with communication in planning. Penny shows how a considered approach can keep dialogue focused and constructive while both capitalising on modern technology and managing expectations. What comes across is a respect and empathy for consultees that makes her approach so effective in bridging the divide between developer and community.'
—Stuart Robinson, Freelance Planning, Placemaking
and Development Consultant and previously
Executive Director and Chairman of Planning at CBRE

'With the increasing demands being placed upon the planning system for local accountability to be at the forefront of planning practice and decision-making, the process of involving the community in planning can frequently be treated as a science. Penny and Martin's book reminds us that engaging with individuals and communities is still very much an art and requires imagination and creativity to both inform and engage effectively. There is a wealth of experience and best practice contained within this book that shows us what has worked and importantly to inspire and encourage us to continue to search for new and innovative ways to engage individuals in the planning and building of our future communities.'
—Will Cousins, Partner, David Lock Associates

'For those involved in promoting Nationally Significant Infrastructure Projects (NSIPs) through the Development Consent Order (DCO) process under the Planning Act 2008 this publication provides a very good account of the stakeholder engagement and public consultation necessary to meet requirements and achieve a successful outcome. Information and guidance on good practice is presented in a clear, concise and informative way which is readily accessible to busy practitioners.'
—Paul White, Technical Director and EIA Technical
Authority in the Infrastructure Division at Atkins Ltd

Public Consultation and Community Involvement in Planning

A Twenty-First Century Guide

Penny Norton
and
Martin Hughes
with a foreword by Louise Brooke-Smith

 Routledge
Taylor & Francis Group

LONDON AND NEW YORK

First published 2018
by Routledge
2 Park Square, Milton Park, Abingdon, Oxon OX14 4RN

and by Routledge
711 Third Avenue, New York, NY 10017

Routledge is an imprint of the Taylor & Francis Group, an informa business

British Library Cataloguing-in-Publication Data
A catalogue record for this book is available from the British Library

Library of Congress Cataloging-in-Publication Data
A catalog record for this book has been requested

ISBN: 978-1-138-68014-2 (hbk)
ISBN: 978-1-138-68015-9 (pbk)
ISBN: 978-1-315-56366-4 (ebk)

Typeset in NewBaskerville
by Apex CoVantage, LLC

Contents

List of figures, images and tables   vii
Foreword   ix
Preface   xi
Acknowledgements   xii

1 Introduction   1

PART I
The context of consultation today   7

2 A brief history of community involvement in planning   9

3 The political climate for community involvement today   24

4 Societal change and community involvement   36

5 The impact of the internet on community involvement   46

PART II
The planning process   67

6 The planning process and the role of consultation   69

7 The formulation of a local plan   81

8 Neighbourhood planning   104

9 Localism and new community rights   121

10 The planning application process   132

11 The role of local authorities in considering and
    determining planning applications   147

12 Appeals and public inquiries   162

13 Consulting on a nationally significant
    infrastructure project   177

PART III
Communications strategy and tactics   191

14 Strategy development   193

15 Tactics to inform and engage   224

16 New consultation tactics   237

17 Analysis, evaluation and feedback   259

18 Reducing risk in consultation   278

PART IV
Post planning   311

19 Community relations during construction   313

20 Community involvement following construction   338

21 Conclusion   355

*Appendix 1: Timeline of political events impacting
on consultation*   358
*Appendix 2: Examples of material and non-material
planning considerations*   362
Appendix 3: Community involvement strategy outline   364
*Appendix 4: Sample content for consultation websites
user guides*   367
Glossary   373
Further reading   389
Index   391

Figures, images and tables

Figures

1.1	Arnstein's Ladder of Citizen Participation	3
5.1	Percentage of daily internet usage by adults	47
5.2	Percentage of internet activities by age group	47
5.3	Percentage of internet activities carried out by UK adults	48
5.4	Percentage of adults aged 16+ in Great Britain accessing the internet away from home or work	49
7.1	Local plan development	86
7.2	Local plan time periods 2009–2015	87
7.3	The sustainability appraisal process	89
14.1	Stages of the strategic process	197
15.1	One- and two-way communication	225
15.2	The nature of consultation tactics	232

Images

15.1	Word cloud showing relative popularity of consultation tactics	224
16.1	Scotch Corner Designer Village consultation website: home page	245
16.2	Scotch Corner Designer Village: registration page	248
16.3	Scotch Corner Designer Village: project video	250
16.4	Scotch Corner Designer Village: document library	251
16.5	Scotch Corner Designer Village: forums landing page	252
16.6	Scotch Corner Designer Village: online forum	252
16.7	Scotch Corner Designer Village Facebook page	253
16.8	Scotch Corner Designer Village Twitter page	254
16.9	VUCITY: an interactive 3D digital model of London	256

17.1   Extract from the Statement of Community Involvement
        for Chelsea Barracks   260
17.2   Visualisation of qualitative research carried out by Fluid
        Design on behalf of Argent (King's Cross) Ltd   262
17.3   The Goodsyard: map showing the geographical location
        of those taking part in the consultation   267
17.4   Crossrail 2: map showing support   268
17.5   Crossrail 2: map showing opposition   269
17.6   Crossrail 2: bar charts showing support/opposition by area   270
19.1   Berkshire House community relations website: home page   330
19.2   Berkshire House: frequently asked questions   331
19.3   Berkshire House: construction updates   331
19.4   Berkshire House: timetable   332

Tables

1.1   Stages of consultation   4
5.1   The focus of campaigns initiated and supported by
       UK community news producers   60
5.2   The focus of investigations carried out by UK community
       news producers   60
5.3   Facebook and Twitter: a comparison   63
8.1   Stages in neighbourhood planning   113
9.1   The Community Right to Build   122
9.2   The Community Right to Challenge   126
9.3   The Community Right to Bid   126
12.1   Comparative success rates in defending planning
       decisions at appeal   172
13.1   The NSIP process   178
14.1   Planning requirements and strategy development   196
14.2   Examples from a typical PEST analysis   198
14.3   Examples from a typical SWOT analysis   198
14.4   Case study: using information to generate messages   208
14.5   Considering anonymity in consultation   211
14.6   NSIPs and the strategic process   216
17.1   Qualitative and quantitative tactics   263

Foreword

> Democracy is a daring concept – a hope that we'll be best governed if all
> of us participate in the act of government. It is meant to be a conversation,
> a place where the intelligence and local knowledge of the electorate sums
> together to arrive at actions that reflect the participation of the largest pos-
> sible number of people.

So said Brian Eno,[1] the renowned musician, producer and songwriter. He is
perhaps not the first person anyone thinks of when considering the English
planning system, but his view of democracy couldn't have been more apt.

As anyone who is involved with the planning regime is aware, while the
approach has remained fairly stable, the systems, regulations and policy
context have been the subject of considerable variation over the years. The
basis of decision-making under our planning regime comprises the sensitive act of encouraging enterprise through the use and development of land
and property, and balancing this with the protection of our environment.

Public participation as part of that regime is increasingly important and
even more so in the wake of 'Localism' and today's 'bottom-up' approach
to planning.

Never before has the involvement of local communities been given so
much support and encouragement, whether it is representing residential,
commercial or industrial areas, or reflecting specific groups within the
community such as LGBT or those of various ethnicities or religions.

The success of those groups or individuals in getting their views across
to the decision makers or indeed those who are pursuing development
proposals is increasingly dependent on factors such as time, funding and
expertise.

In an ideal world, the process of consultation needs to be explained to
the whole community and involvement specifically invited. The better the
communication of proposals, the better the involvement of the community.
While individuals commonly speak up if they are opposed to local proposals, rarely do they comment if they generally support a new scheme – but
does this means that the quantity of comments should hold sway, as opposed

to the quality of those comments? Groups, however, can act in a different way, and the art of the 'pressure group' is a phenomenon that the planning system is learning to cope with.

This book is an excellent exposé of the science and art of consultation and public involvement in the planning system, looking at how this influences the way development proposals are prepared, pursued and eventually determined.

As Brian Eno eloquently suggested, democracy and participation can be very daring concepts but if managed and resourced appropriately, they can help in bringing forward schemes and proposals that reflect the wishes and aspirations of all players, be they members of the local community, the developer, the financier or elected members.

The art of involvement and community liaison is the bedrock of today's planning system and when it works well, it reflects the very best in terms of inclusivity. When managed poorly or not given the respect it deserves, it can result in poor decision-making and at worst, some very dodgy development coming forward.

So I very much welcome this book in the way it assesses good and bad practice from across the country and fully expect it to become a source of guidance and support for all those who want to see an effective and successful planning regime.

Louise Brooke-Smith
Director, Brooke Smith Planning Consultants
RICS President 2014–2015

Note

1 *The Guardian*, 8 July 2009. Available www.theguardian.com/commentisfree/2009/jul/08/government-electoral-process-voters

Preface

The twenty-first century has seen significant changes in consultation and community involvement in planning.

Changes to the environment in which community involvement exists include an increased legal requirement to consult; a renewed focus on engagement through 'Localism' and other legislative measures; changes to the way in which we, as a society, define ourselves; and advances in technology which enable communities to organise, communicate and respond to development proposals quickly and effectively.

This book assesses the impact of these changes and provides practical advice as to how professionals – property developers and planners, local authorities, the infrastructure and energy sectors – can run effective consultations today.

It also embraces the opportunities posed by new methods of consultation. A rise in participative initiatives provides effective, 'two-way' consultation and more meaningful, qualitative responses. And as the revolution in online communication has resulted in the majority of development proposals (whether intended by the applicant or otherwise) having an online profile, it addresses the benefits of online consultation.

The first section of the book (Chapters 2–5) provides a context for community involvement today. The second section (Chapters 6–13), by Martin Hughes, is a discursive view of the process of consultation within the planning systems, and the third section (Chapters 14–18) looks at the strategy and tactics of consultation. Finally, Chapters 19 and 20 address the continuing role of community involvement both during construction and thereafter.

I am extremely grateful to a significant number of people who have provided insight through interviews, comment or case studies, resulting in a book which brings together some of the best practice within consultation, community relations and community involvement – from the first planning meeting through to construction and beyond.

I hope that this book will provide both a useful resource and an inspiration.

Penny Norton
2016

Acknowledgements

The authors would like to thank the following for sharing their knowledge and insight: David Alcock (Anthony Collins Solicitors LLP), James Anderson (Turley), Shabana Anwar (Bircham Dyson Bell LLP), Nick Bailey (University of Westminster), Nick Belsten (Indigo Planning), Jonathan Bradley (Participate), Peter Bradley (Transport for London), Louise Brooke-Smith (Brooke Smith Planning Consultants Ltd), Mark Brookes (Dacorum Borough Council), Keith Butterick (University of Huddersfield), Will Cousins (David Lock Associates), Hilary Cox (community artist), Gary Day (McCarthy & Stone), Emma Fletcher (SmithsonHill Estates Limited), Simon Ford (Amec Environment and Infrastructure UK), Jon Grantham (Land Use Consultants Limited), Ron Henry (Peter Brett Associates), Nick Keable (Development Intelligence), Lyn Kesterton (Locality), Andy Lawson (Gallagher Estates), Rachel Lopata (Community Research), Roger Madelin CBE (Argent LLP), Tracy Mann (freelance facilitator), Steve McAdam (Soundings), Jacqueline Mulliner (Terence O'Rourke), Adrian Penfold (British Land plc), Pauline Roberts (Nathaniel Lichfield & Partners), Carole Robertson (Ferring District Council), Rebecca Saunt (East Cambridgeshire District Council), Alan Smith (Kier Group plc), Malcolm Smith (Arup), Carl Taylor (Redditch Co-operative Homes and Ashram Moseley Housing Association), Andy Thompson (Beacon Planning), Stuart Thomson (Bircham Dyson Bell LLP), Heather Topel (North West Cambridge), Alison Turnbull (Alison Turnbull Associates), Malcolm Wagstaff (Cringleford Parish Council), Biky Wan (North West Cambridge), Ben White (Crossrail), Nicola White (Arup), Paul White (Atkins Global), Debbie Wildridge (Foundation East Limited) and Nick Woolley (Woolley).

Special thanks to Chris Quinsee and Ruth Child of Q+A Planning, and Judith Mack and Phil Kennedy for their technical and editorial expertise.

1 Introduction

Why involve the public in development?

In 2011, embarking on the Localism agenda which was to set the scene for community involvement in planning today, the government said:

'Pre-application consultation provides an opportunity to achieve early consensus on controversial issues before proposals are finalised. This should encourage greater community engagement in the process, and result in better quality applications submitted to local authorities, which are more in line with community aspirations, and much less controversial. Such an approach is considered to be inclusive and transparent, with development outcomes more in line with what the community desires.'[1]

The statement was based on the following assumptions:[2]

- As a result of pre-application consultations, there is a 10–15% fall in the number of appeals, hearings and enquiries.
- Greater community involvement will benefit society by providing a positive and constructive role for local people in the planning process. The resultant increase in local support for new development should lead to more, better quality housing (and other development) being delivered.
- Many of the 10–20% of developers who do not currently consult do so not because there are no benefits of consulting, but because they (unlike their industry colleagues) have not caught on to the benefits.

Localism was largely driven by the need to substantially increase the UK's housing stock by expediting the rate at which planning permission is granted, and it was based on the belief that local involvement would deliver greater consensus. Whether Localism has ultimately achieved its aims is a subject of much debate, but the notion that community involvement *can* benefit planning decisions is unequivocal.

Planning is ultimately about people: whether a local authority-run strategic plan or a private sector-led development proposal, change to the built environment impacts on communities. It is generally believed that those proposing changes should involve local residents as a courtesy, but planners and developers also have much to benefit from involving local people.

Consultation provides the opportunity for the development team to glean information and ideas from the local community. This might include knowledge of local history and which has the potential to enrich a scheme, otherwise unknown social issues which might have delayed the process, and the needs and aspirations of the community which may be met through the new development. With local input, proposals can be enriched and finely tuned to a specific neighbourhood, creating a unique scheme well suited to its location.

The local community, too, can benefit: community involvement can promote social cohesion, strengthen individual groups within it and create a shared legacy.

Having consulted thoroughly at an early stage and having had proposals either validated or challenged, a developer has a greater chance of building localised consensus and support for a proposed scheme.

A well-run consultation can build a trusting and mutually cooperative relationship between the consulting body and the community, which can minimise the potential for conflict and thereby remove the risk in the process. A widely targeted consultation can also increase the likelihood of support: those living closest to any proposed development are traditionally the most likely to object, but those further afield may welcome the additional facilities without concern about the construction process.

The proposed expansion of Heathrow Airport has attracted significant community opposition over many years. The Royal Town Planning Institute (RTPI) drew attention to this in 2008, highlighting that a 'growing militancy of protests' was emerging as a direct result of 'consultation processes which undervalue community input', and that 'failure to reform the system could create a new generation of Swampy-style protestors'.[3]

It said,

> The RTPI believes the Heathrow protests, which resulted in security breaches at both Heathrow Airport and the Houses of Parliament, occurred because the protestors believe they are powerless to effect change through official channels. It warned that unless the community consultation process is significantly improved, high profile protest activity against major infrastructure developments will become common once again.

This example shows how the consultation process, if run badly or half-heartedly, can have a negative impact. But the reverse is true: a good consultation can both strengthen communities and benefit development proposals. In addition, it has the potential to enhance the reputation of the consulting body and in doing so can also benefit community relations during construction and beyond.

Defining community involvement

The terms *consultation, community engagement* and *community involvement* are regularly and interchangeably used.

Throughout this book, we use the terms *community engagement* and *community involvement* in relation to the wide-ranging communication which may take place between either a local authority and its residents, or a developer and the neighbours or future users of a potential development.

More specifically, the term *consultation* is used to describe the process whereby an organisation communicates with the public for the purpose of shaping a development plan or planning application.

Communications theory

In 1969, the US communications academic Sherry Arnstein identified the terminology of consultation in her Ladder of Citizen Participation.[4] Although our usage of the terms differs, Arnstein's theory is a useful resource by which community involvement can be considered today.

Arnstein's ladder (Figure 1.1) was intended to reflect the relationship between community and government, identifying poorly led participation as 'manipulation' on the bottom rung of the ladder and rising to 'citizen control' at the very top. Although communications professionals today take issue

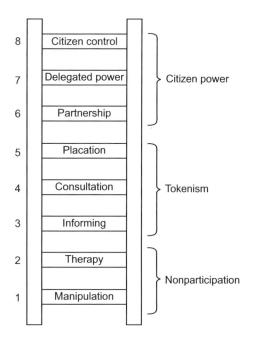

Figure 1.1 Arnstein's Ladder of Citizen Participation

with 'consultation' being termed 'tokenism', and most would quite justifiably choose to ignore the top two rungs of the ladder, the framework has endured because of the way in which it so clearly recognises varying levels of community involvement.

We would suggest alternating the positions of consultation and placation, thereby positioning consultation at the centre of the ladder, representing a process which involves local people and government/organisations equally. Ideally the fifth rung would be divided further to reflect the fact that consultation itself has many forms, as is shown in Table 1.1:

Table 1.1 Stages of consultation

8	**Citizen control** Communities develop land without planning consent	Citizen power
7	**Delegated power** Community self-build	
6	**Partnership** Developer instigates a Planning for Real® exercise to involve residents in the early development of proposals	Involvement
5	**Involvement** Developer encourages comments on proposals in focus groups, meetings and online forums	
4	**Information** Developer communicates information about a proposed scheme without a means of feedback	Tokenism
3	**Placation** Developer communicates only positive information about a proposed scheme without a means of feedback	
2	**Therapy** Developer uses subtle means to discourage residents from objecting to a proposed scheme	Non-participation
1	**Manipulation** Developer actively encourages residents to support a scheme or discourages them from objecting to it	

Stages in community involvement

Community involvement is not a brief break in bureaucratic proceedings when a developer or local authority asks a local community to approve its proposals.

Effective community involvement is a long-term process. At a strategic planning level, the formulation of a local plan has many stages at which residents may contribute, and simultaneously residents have the opportunity to become involved in land-use decisions by taking part in (or even leading) neighbourhood plan consultations. For a specific development, local involvement can include input into the first scoping document, consideration of the options, the drawing up of detailed plans, contributing to the local authority's consultation, and remaining involved in communication with the developer during construction and beyond.

Conclusion

This book provides a companion to each stage of community involvement, from the formulation of a local plan or neighbourhood plan through to community-building post-construction. It is intended to be read by local authority planners, planning consultants, developers, communications professionals – indeed all those involved in the community involvement process. Providing an overview of the recent changes affecting community involvement (Chapters 2–5), a discursive look at the process of planning (Chapters 6–13), and practical advice and best practice (Chapters 14–20), it aims to inform, provoke and advise, and in doing so, fully prepare its readers to oversee community interaction effectively.

Notes

1 Department for Communities and Local Government (2011) Localism Bill: Compulsory Pre-Application Consultations Between Prospective Developers and Local Communities Impact Assessment. London: DCLG.
2 Ibid.
3 RTPI (2008) *PR 11 – Heathrow Protests May Be Just the Beginning* [Online]. Available www.rtpi.org.uk/briefing-room/news-releases/2008/february/pr-11-heathrow-protests-may-be-just-the-beginning-rtpi [Accessed 12 October 2016]
4 Participatory Methods (2014) *The Ladder of Citizen Participation* [Online]. Available www.participatorymethods.org/method/levels-participation [Accessed 12 October 2016]

Part I

The context of consultation today

2   A brief history of community involvement in planning

The emergence of community involvement in planning: 1945–1979

The conclusion of the Second World War brought about many radical changes, and planning was not exempt. In fact the 1947 Planning Act is widely regarded as one of the most significant pieces of legislation in planning history.

The Act firmly established the requirement for planning permission, removing the centuries-old rights of landowners to develop land as they chose. It established 145 new planning authorities (borough and county councils), each of which was required to prepare a comprehensive development plan. In doing so, and in bestowing on local authorities the right to carry out redevelopment of land themselves, the 1947 act introduced two central pillars of today's planning system: those of strategic planning and public sector-led approval of development.

Although the Act did not introduce a requirement to consult with the public, it transferred development decisions in both strategic planning and specific planning applications from powerful landowners to local authorities, therefore put planning into the hands of democratically elected politicians.

Within the next two decades, the concept of community involvement became more widely understood, albeit in sectors other than planning. Driven by global organisations such as the UN, the World Health Organisation (WHO) and UNICEF, and inevitably the liberating spirit of the 1960s, community participation emerged and flourished. Community design centres emerged in the US in the 1960s and not surprisingly pressure to increase community involvement in planning followed in the UK.

The 1968 Town and Country Planning Act introduced structure plans, the first form of strategic level development plans. Prepared either by a county council or by local authorities working jointly together, structure plans created a long-term vision for land use in a specific area with regard to major land uses such as housing, education, recreation, and transport.

Local development plans, prepared by district rather than county councils, were required to accord with the overall strategy set out in the structure plan. While the former required government approval, the latter continued to be determined and approved by local planning authorities.

By 1968 it was widely acknowledged that community participation was lacking in UK planning, and the Labour government appointed Arthur Skeffington, then parliamentary private secretary to the minister for housing and local government, to assess how the 'top-down' system inherent in the Town and Country Planning Act might be rectified and public participation in planning increased. Widely known simply as the 'Skeffington Report', *People and Planning* was published in 1969. The report defined 'participation' as 'The act of sharing in the formulation of policies and procedures'[1] and proposed that local development plans be subject to full public scrutiny and debate. It also identified the role of conservation societies in consultation, recommending that such groups were regarded as central to any public consultation, while also acknowledging that amenity groups alone were not representative of a community.

Despite its positive reception, few of the report's recommendations were put into practice. It has been suggested that they were too vague and intangible. That said, the Skeffington Report had an enduring influence on the way in which politicians view involvement on planning issues – to the extent that Skeffington is said to have influenced the introduction of Localism over 40 years later.

Eight years after the publication of the Skeffington Report, a report by the Town and Country Planning Association, *The Crisis in Planning* (1977)[2], expressed how little had changed in relation to public involvement in planning. The report said that public disillusionment with the planning process was so 'widespread that one does not feel obliged to document it', and that planners were seen as 'narrowly defined technical experts in administering state regulations and managing aspects of the land development process'.

Business-driven planning: 1979–1997

Political involvement in planning – together with politicians' attitudes towards public participation – shifted considerably under the Conservative governments of the 1980s and early 1990s. Privatisation impacted on all aspects of public life, adversely affecting public participation in planning as the influence of local authorities was gradually eroded – and in the case of the Greater London Authority and metropolitan district councils, removed entirely. While the new-found entrepreneurial spirit led to substantial new development schemes in key areas such as the London Docklands, the establishment of urban development corporations centralised the power of the planning system in authorising development, and thereby reduced the role of local communities in participating in decisions.

The title of the Conservative government's 1985 white paper, *Lifting the Burden*[3], summarises the administration's attitude towards the planning regime that it inherited. The white paper was part of a continuing programme to reduce bureaucracy. Specifically, it aimed to lift 'the administrative and legislative burdens – one of which was planning controls which take time, energy and resources from fundamental business activity'.

The Conservative government's first major piece of planning legislation was eleven years after gaining control. The 1990 Town and Country Planning Act clarified the two distinct functions of planning that we are familiar with today: forward planning and development control. It also updated planning policy across a range of areas, including the nature and structure of planning authorities, strategic planning responsibilities, and the procedures by which planning applications are determined.

Perhaps most significantly, the 1990 Town and Country Planning Act put in place planning obligation agreements, more commonly known today as Section 106 agreements. Section 106 requires that developers make a substantial financial contribution to mitigate any negative impact of development on the local community. Although largely replaced by the community infrastructure levy in 2004, Section 106 remains in some circumstances and often provides an opportunity for a developer to make a positive contribution to a local community and build a relationship which can be of benefit during the construction phase and beyond.

During the period 1992 to 2012, a series of Planning Policy Statements (PPSs) were produced to replace the previous Planning Policy Guidance notes (PPGs). Like their predecessor, PPSs were guidance rather than law, but all local authorities were required to take the guidance into consideration when drawing up development plans or making specific decisions on planning applications. If, therefore, a local authority had overlooked the guidance in the case of a strategic planning document or planning application which was the subject of an inquiry, the fact that government policy had been disregarded would be a major consideration. PPSs covered a range of topics, including delivering sustainable development, housing, planning for sustainable economic growth, planning for the historic environment, planning for town centres, sustainable development in rural areas, planning for sustainable waste management, local development frameworks and renewable energy. From a community involvement point of view, the most significant was Planning Policy Guidance (PPG) Note 1, General Policies and Principles (1997), which was later replaced by Planning Policy Statement 1: Delivering Sustainable Development (2005). Despite being amended through their twenty-year existence, PPSs remained in place until the publication of the National Planning Policy Framework in 2012.

Box 2.1 Planning policy statement 1: delivering sustainable development[4]

PPS 1 required that the following key principles be applied to ensure that development plans and decisions taken on planning applications contribute to the delivery of sustainable development:

1   Development plans should ensure that sustainable development is pursued in an integrated manner, in line with the principles for sustainable development set out in the UK strategy. Regional planning bodies and local planning authorities should ensure that development plans promote outcomes in which environmental, economic and social objectives are achieved together over time.

2   Regional planning bodies and local planning authorities should ensure that development plans contribute to global sustainability by addressing the causes and potential impacts of climate change – through policies which reduce energy use, reduce emissions (for example, by encouraging patterns of development which reduce the need to travel by private car, or reduce the impact of moving freight), promote the development of renewable energy resources, and take climate change impacts into account in the location and design of development.

3   A spatial planning approach should be at the heart of planning for sustainable development.

4   Planning policies should promote high quality inclusive design in the layout of new developments and individual buildings in terms of function and impact, not just for the short term but over the lifetime of the development. Design which fails to take the opportunities available for improving the character and quality of an area should not be accepted.

5   Development plans should also contain clear, comprehensive and inclusive access policies – in terms of both location and external physical access. Such policies should consider people's diverse needs and aim to break down unnecessary barriers and exclusions in a manner that benefits the entire community.

6   Community involvement is an essential element in delivering sustainable development and creating sustainable and safe communities. In developing the vision for their areas, planning authorities should ensure that communities are able to contribute to ideas about how that vision can be achieved, have the opportunity to participate in the process of drawing up the vision, strategy and specific plan policies, and to be involved in development proposals.

Box 2.2 The Nolan Principles (1995)

The Committee on Standards in Public Life, led by Lord Nolan, was an advisory non-departmental public body set up in 1994 to advise the prime minister, John Major, on ethical standards of public life following a series of embarrassing incidents. Although not specifically related to planning, The Seven Principles of Public Life identified in Nolan's 1995 report have become established best practice in local government and as such are often referred to in the determination of planning applications.

1 Selflessness

   Holders of public office should act solely in terms of the public interest.

2 Integrity

   Holders of public office must avoid placing themselves under any obligation to people or organisations that might try inappropriately to influence them in their work. They should not act or take decisions in order to gain financial or other material benefits for themselves, their family, or their friends. They must declare and resolve any interests and relationships.

3 Objectivity

   Holders of public office must act and take decisions impartially, fairly and on merit, using the best evidence and without discrimination or bias.

4 Accountability

   Holders of public office are accountable to the public for their decisions and actions and must submit themselves to the scrutiny necessary to ensure this.

5 Openness

   Holders of public office should act and take decisions in an open and transparent manner. Information should not be withheld from the public unless there are clear and lawful reasons for so doing.

6 Honesty

   Holders of public office should be truthful.

7 Leadership

   Holders of public office should exhibit these principles in their own behaviour. They should actively promote and robustly support the principles and be willing to challenge poor behaviour wherever it occurs.

New Labour and a commitment to involve: 1997–2010

Labour's momentous electoral success in 1997 after 18 years in the political wilderness marked a huge ideological shift in attitudes towards community involvement in planning.

Over the ensuing 13 years of Labour administration, collaboration and public involvement played a central role, albeit with varying levels of success, across all area of public life. Government initiatives impacting on planning included the National Strategy for Neighbourhood Renewal, local strategic partnerships, Community Engagement Network, Local Agenda 21 and the New Deal for Communities.

The philosophical background for the government's new public sector ideology is eloquently expressed by Anthony Giddens,[5] who argued that society needed new 'forms of democracy other than the orthodox voting process', and called for 'governments to re-establish more direct contact with citizens and citizens with government through "experiments with democracy, electronic referenda, citizens' juries and other possibilities"'.

The 1998 white paper, *Modern Local Government, In Touch with the People,*[6] was an inevitable extension of this thinking. It contained four main proposals which addressed the perceived failings in democratic practice:

- Electoral reform – a number of measures were proposed to enhance and increase participation with the overall aim of raising voter turnout at local and national elections. This included voting on different days, experimenting with electronic voting systems and introducing some form of proportional representation in elections.
- Improved management of the political process. This produced the most tangible political change, replacing the long-standing committee-based system with a council leader, cabinet structure and Westminster-style scrutiny committees, and where the local population demanded one in a referendum, directly elected mayors.
- Extending local autonomy and community leadership. Local authorities were to promote the 'economic, social and environmental well-being of their areas'.
- Encouraging and improving public participation. Public participation in local government was to be enhanced by using a range of consultation and participation processes.

Specifically, the white paper stated, 'The Government wishes to see consultation and participation embedded into the culture of all councils, including parishes, and undertaken across a wide range of each council's responsibilities'.

This was an admirable, but unachievable, objective according to some. Needham (2002)[7] commented that engagement, 'Has been presented as the panacea for all that ails local democracy, building trust between people and their elected representatives, whilst ensuring that authorities are accountable and responsible to their communities', and argued that by 2002, 'Consultation has not so far achieved the reinvigoration of local democracy that had been expected'.

The white paper was followed by several reviews, policy documents and pieces of legislation which concerned community involvement:

- The 1999 Local Government Act included a requirement for all local authorities to consult with taxpayers and statutory guidance emphasised the need to consult widely while undertaking Best Value reviews.
- The 2000 Local Government Act required local authorities to prepare community strategies to promote economic, social and environmental well-being and bring about sustainable development.
- The theme of increasing public involvement was continued in 2001 with the white paper *Strong Local Leadership – Quality Public Services*, which stated, 'We will support councils in their efforts to lead their communities and meet people's needs. In particular we will support greater levels of community engagement and involvement in council business'.
- In 2004 the Office of the Deputy Prime Minister (ODPM) produced a substantial document, *Community Involvement in Planning*, which explained the government's commitment to planning in advance of the 2004 Planning and Compulsory Purchase Act.
- The Gershon Report of 2004, *Releasing Resources from the Front Line*, was an independent review of public sector efficiency, published by Sir Peter Gershon, and largely concerned the debate about public services and the role of government.
- The paper *Citizen Engagement and Public Services: Why Neighbourhoods Matter*, published by the ODPM in 2005, stated, 'Government departments have adopted a common framework for building community capacity and agreed a shared objective: to increase voluntary and community engagement, especially amongst those at risk of social exclusion, and increasing the voluntary and community sector's contribution to delivering public services'.
- In 2006, a white paper, *Strong & Prosperous Communities*, established a further change in direction for local government, with a heavy emphasis on devolving power and increased public participation and engagement.

Box 2.3 Aarhus Convention 1998

The United Nations' Economic Commission for Europe (UNECE) Convention on Access to Information, Public Participation in Decision-making and Access to Justice in Environmental Matters was signed on 25 June 1998 in the Danish city of Aarhus. Its signatories include most EU members and include the UK.

The Aarhus Convention grants the public rights regarding access to information, public participation and access to justice, in governmental decision-making processes on matters concerning the local, national and transboundary environment. As such it has become an important benchmark in consultation, specifically in relation to dialogue between the public and public authorities.

The Regional Development Agencies Act 1998 led to the formation of eight regional development agencies (RDAs)[8] and the London Development Agency (LDA). Their purpose was to further economic development and regeneration; to promote business efficiency and competitiveness; to promote employment and enhance the development and application of skills relevant to employment; and to contribute to sustainable development. Each RDA was led by a chair and a board of fifteen people, appointed by a minister of the Department for Business, Innovation and Skills. The RDA chairs were drawn from the business sector, and the boards represented business, local government, trade unions and voluntary organisations.

Box 2.4 The Freedom of Information Act 2000

The Freedom of Information Act (FOIA), which took effect in January 2005, was put in place to ensure that everyone had a right of access to information from public authorities. The Act also requires public authorities to have regard to the public interest in allowing access to information and in publishing reasons for decisions.

The FOIA covers any recorded information that is held by public authorities, including government departments, local authorities, the NHS, state schools and police forces.

In planning, the FOIA is a powerful tool enabling individuals to request information from a local authority in relation to a planning decision and to potentially use that information when challenging the local authority's decision. Comment from members of the public on planning applications is available through the FOIA, although personal data may not be disclosed. Campaigners or those questioning a planning committee's decision may also use the FOIA to access local authority correspondence with other parties. Information about decision makers or influencers may be sought, which is particularly relevant when it is believed that an individual's position on a planning decision may be tainted by a conflict of interest.

Perhaps the most significant piece of legislation introduced by the Labour government during the early twenty-first century was the 2004 Planning and Compulsory Purchase Act. The Act sought to speed up the planning system, introduce greater transparency in decision-making and reaffirm sustainable development as an over-riding priority.

Building on the structures put in place by the Regional Development Agencies Act 1998, the Planning and Compulsory Purchase Act abolished the structure plans which had been the bedrock of strategic planning since 1968. In their place, it introduced statutory regional planning. Regional planning bodies known as regional assemblies (RAs) were tasked with preparing regional spatial strategies (RSSs) for the eight English regions.

RSSs were expected provide an integrated and strategic spatial framework for a region for a 15- to 20-year period and to:

- Establish a 'spatial' vision and strategy for the region
- Contribute to sustainable development
- Address regional or sub-regional issues that might have crossed county, unitary authorities or district boundaries
- Produce housing targets
- Produce a regional waste strategy
- Outline infrastructure investment priorities
- Outline key priorities for investment, particularly in infrastructure

The Act also replaced old-style local plans and unitary development plans with local development frameworks (LDFs), which were intended to contribute to the achievement of sustainable development within the policies set out in the RSSs. Consequently, critics accused the Labour government of reducing local involvement in planning through distant, regionally based decision-making.

From a consultation point of view, the introduction of the Statement of Community Involvement (SCI) was perhaps the most significant innovation in the 2004 act, and may be the most enduring of its entire content.

Section 18 of the Act requires that each local planning authority put in place an SCI to establish how it intends to involve the public in creating planning policy documents and consulting on planning applications. It is required to convey the means by which the council will inform, consult and involve the community, and explain the means by which local people might get involved. In some cases the SCI will detail the communities and organisations in the area, including information about traditionally 'hard-to-reach' groups. The planning authority should set out the steps that they will take to involve such groups.

Box 2.5 Statement of Community Involvement: typical content

Introduction

- The planning system – national and local context
- What is the Statement of Community Involvement?
- Why is the Statement of Community Involvement needed?
- Principles of Community Involvement
- Methods for Community Involvement
- Sustainability appraisal
- Monitoring and reviewing community involvement

Consultation on the local plan

- The local plan process
- Community involvement methods
- Timescales
- Audiences
- Notification methods
- Feedback

Community involvement in planning applications

- The planning application process
- Pre-application consultation
- Landowner and developer interests
- Commenting on planning applications
- Notification methods
- Appeals
- Feedback

Neighbourhood planning

- The neighbourhood planning process
- Community involvement opportunities

- Neighbourhood plans in existence
- Opportunities to get involved

Consultation on the community infrastructure levy

- The local authority's CIL schedule
- Opportunities to comment on CIL

Appendixes

- List of potential consultees (including hard-to-reach groups)
- Consultation bodies
- Material and non-material considerations
- Complaints procedure
- Key contacts
- Further information and resources
- Glossary

Unlike many of the planning policies introduced by the Labour governments of 1997–2010, SCIs are still in place today. For local authorities they are the foundations upon which all consultations are built, and for developers they are an important starting point in understanding a particular local authority's attitude towards consultation. More importantly still, they provide developers with an understanding of what is judged by the local authority to be acceptable in consultation, as well as a useful resource in understanding the groups which operate within the community.

Contrary to the principles of devolving powers which had been a feature of many of the Labour governments' white papers and, to an extent, legislation, the 2008 Planning Act created a system for approving the construction of major infrastructure projects via a new body, the Infrastructure Planning Commission (IPC) and its 35 appointed commissioners. The public inquiry system was replaced with an open floor hearing, but the public had no right to bring witnesses or conduct cross-examinations. National policy statements (NPS) were introduced as part of the Act. They established government policy on major infrastructure projects such as airports, power stations, major roads, railways, ports, reservoirs and hazardous waste facilities.[9]

Importantly, the Act also made provision for the introduction of a community infrastructure levy, which largely replaced Section 106 payments and remains in place today.

Coalition politics and devolved powers: 2010–2015

On 5 May 2010, the country elected its first Conservative prime minister in 13 years, albeit as head of a coalition with the Liberal Democrat party.

The new 'compassionate conservatism' had little in common with that of Thatcherism: its first major policy announcement concerned the restoration of power to local authorities and a raft of community-centred initiatives.

Launching the 'Big Society' just two weeks after the general election, the new Prime Minister David Cameron said,

> What is my mission? It is actually social recovery . . . to mend the broken society that is what the Big Society is all about . . . responsibility is the absolute key, giving people more control to improve their lives and their communities, so people can actually do more and take more power. It is a different way of governing . . . But above all, it's entrepreneurship that is going to make this agenda work.[10]

While many critics of the new government attributed the sharing sentiment to that of the Coalition, the origins of the Big Society can be seen clearly in a Conservative Party green paper from June 2008, entitled *A Stronger Society: Voluntary Action for the 21st Century.*

Box 2.6 Building the Big Society: a policy paper published on 18 May 2010 by the Conservative and Liberal Democrat coalition government

Our Conservative-Liberal Democrat Government has come together with a driving ambition: to put more power and opportunity into people's hands. We want to give citizens, communities and local government the power and information they need to come together, solve the problems they face and build the Britain they want. We want society – the families, networks, neighbourhoods and communities that form the fabric of so much of our everyday lives – to be bigger and stronger than ever before. Only when people and communities are given more power and take more responsibility can we achieve fairness and opportunity for all. Building this Big Society isn't just the responsibility of just one or two departments. It is the responsibility of every department of Government, and the responsibility of every citizen too. Government on its own cannot fix every problem. We are all in this together. We need to draw on the skills and expertise of people across the country as we respond to the social, political and economic challenges Britain faces. This document outlines the already agreed policies that we believe will help make that possible. It is the first strand of a comprehensive Programme for Government to be published in the coming days, which will deliver the reform, renewal, fairness and change Britain needs.

1   Give communities more powers

  - We will radically reform the planning system to give neighbourhoods far more ability to determine the shape of the places in which their inhabitants live.
  - We will introduce new powers to help communities save local facilities and services threatened with closure, and give communities the right to bid to take over local state-run services.
  - We will train a new generation of community organisers and support the creation of neighbourhood groups across the UK, especially in the most deprived areas.

2   Encourage people to take an active role in their communities

  - We will take a range of measures to encourage volunteering and involvement in social action, including launching a national 'Big Society Day' and making regular community involvement a key element of civil service staff appraisals.
  - We will take a range of measures to encourage charitable giving and philanthropy.
  - We will introduce a National Citizen Service. The initial flagship project will provide a programme for 16 year olds to give them a chance to develop the skills needed to be active and responsible citizens, mix with people from different backgrounds, and start getting involved in their communities.

3   Transfer power from central to local government

  - We will promote the radical devolution of power and greater financial autonomy to local government, including a full review of local government finance.
  - We will give councils a general power of competence.
  - We will abolish Regional Spatial Strategies and return decision-making powers on housing and planning to local councils.

4   Support co-ops, mutuals, charities and social enterprises

  - We will support the creation and expansion of mutuals, cooperatives, charities and social enterprises, and support these groups to have much greater involvement in the running of public services.
  - We will give public sector workers a new right to form employee-owned cooperatives and bid to take over the services they deliver. This will empower millions of public sector workers to become their own boss and help them to deliver better services.

- We will use funds from dormant bank accounts to establish a Big Society Bank, which will provide new finance for neighbourhood groups, charities, social enterprises and other non-governmental bodies.

5   Publish government data

- We will create a new 'right to data' so that government-held datasets can be requested and used by the public, and then published on a regular basis.
- We will oblige the police to publish detailed local crime data statistics every month, so the public can get proper information about crime in their neighbourhoods and hold the police to account for their performance.

The devolution of power to communities and local government was a central plank of the new policy, providing greater autonomy over finances, housing, and planning to local government. The policy also set out some significant community-focused initiatives, including the creation of an independent 'Big Society Bank', initiatives to encourage greater charitable giving and philanthropy, support for the creation and expansion of mutuals, cooperatives, charities and social enterprises, the introduction of a National Citizen Service for 16-year-olds and the creation of a new 'right to data' that allows government-held data to be requested and used by the public and will oblige the police to publish detailed local crime statistics every month.

Importantly, from a community involvement point of view, this declaration of intent from the new administration established the principles of Localism, which was to take planning policy in another new direction.

Notes

1 Child, P. (2015) People and Planning: Report of the Committee on Public Participation in Planning (the Skeffington Committee Report) With an Introduction. *Planning Perspectives*, 30(3), 484–485.
2 Town and Country Planning Association (1977) *The Crisis in Planning*. London: TCPA.
3 The Conservative Party (1985) *Lifting the Burden*. London: The Conservative Party.
4 Office of the Deputy Prime Minister (2005) Planning Policy Statement 1: Delivering Sustainable Development. London: ODPM.
5 Giddens, A. (1998) *The Third Way, the Renewal of Social Democracy*. Cambridge: Polity Press.
6 Department of the Environment, Transport and the Regions (DETR) (1998) *Modern Local Government, in Touch With the People*. London: DETR.

7 Needham, C. (2002) Consultation: A Cure for Local Government? *Parliamentary Affairs*, 55, 699–714.
8 The individual Regional Development Agencies were:

- East of England Development Agency (EEDA), based in Cambridge
- East Midlands Development Agency (EMDA), based in Nottingham
- London Development Agency (LDA)
- One NorthEast (ONE), based in Newcastle
- Northwest Regional Development Agency, based in Warrington
- South West of England Regional Development Agency, based in Exeter
- South East England Development Agency (SEEDA), based in Guildford
- Advantage West Midlands, based in Birmingham
- Yorkshire Forward, based in Leeds

9 The history and current procedures in relation to NSIPs are described in detail in Chapter 13.
10 Cameron, D. (2011) *PM's Speech on Big Society [Online]*. Available www.number10.gov.uk/news/pms-speech-on-big-society

3 The political climate for community involvement today

Introduction

The previous chapter focused on the history of community involvement in planning from its instigation in 1947 to the changed political backdrop from 2010, in particular identifying initiatives and major pieces of legislation which have created a lasting impact. This chapter creates a picture of the political climate for community involvement in planning today – be it as a result of legislation (both historic and more recent) or simply political influences and attitudes.

A political imperative which underlines all planning policy of the twenty-first century, and has become an increasingly high priority, is the need to develop more new homes, at a faster rate than ever before. There is cross-party consensus on this view. Consequently, initiatives to speed up the planning system, with a focus on residential development, have been a feature of all recent legislation.

Localism (2011)

Localism was a central plank of Coalition (2010–2015) and (arguably to a lesser extent) later Conservative policy and as such is the backbone of much of our current planning law.

The Localism Act was published in November 2011 and delivered on many of these sentiments. It contained a wide range of measures to devolve powers to councils and neighbourhoods and give local communities greater control over local decisions. The Act contained four main provisions:

- New freedoms and flexibilities for local government
- New rights for communities and individuals
- Reform to make the planning system more democratic and effective
- Reform to ensure that decisions about housing are taken locally.[1]

As we shall see in Chapter 8, neighbourhood planning was perhaps the most significant output of Localism. Other initiatives, which are covered in greater detail in Chapter 9, are a series of community rights, including the Community Right to Build, Community Right to Challenge, Community Right to Bid, Community Right to Contest and the Community Asset Transfer.

Localism is also present in a number of other policies introduced in the early days of the Coalition government and still relevant today. These include the new homes bonus, the local government resource review, tax increment finance and renewable energy projects.

In 2011, the *Open Public Services* white paper addressed a move from 'top-down' to 'bottom-up' thinking, based on the principles of greater choice, decentralisation, more diversity in service providers, fair access and accountability in local government. The policy shift reflected traditional Conservative policies in its aspiration to open up the commissioning of public services and a commitment to intervene if local services fail, but also that of the new 'compassionate conservatism' and the influence of Liberal Democrats in government in its commitment to put communities and neighbourhoods at the heart of local services. The white paper emphasised the importance that 'Voice comes through participation in service design or management, and via champions in the form of elected representatives such as councillors and unelected representative bodies such as consumer organisations.'[2] It also stated that 'There will be an increase in web-based services that allow consumers of individual public services to share opinions and to compare performance data.'

Predetermination

For developers wishing to involve with councillors on development proposals, the change in law concerning predetermination is of benefit. Previously, developers were discouraged from communicating with planning committee members in advance of a decision being taken by the committee, as it was believed that councillors would make up their minds in advance, one way or other, and therefore not engage in decision-making in an open and considered way. The Localism Act clarified the rules on predetermination, establishing that councillors may play an active part in local discussions without fear of legal repercussions.

A requirement to consult

Perhaps most importantly from a community involvement point of view, to further strengthen the role of local communities in the planning process, the Localism Bill introduced a new requirement for developers to consult local communities before submitting planning applications for certain developments, the intention being that this would give local people a chance to comment when there was still genuine scope to make changes to proposals.

In March 2010, a poll carried out by YouGov for the National Housing and Planning Advice Unit had demonstrated that 21% of respondents opposed new housing supply in their area, but that this number fell to 8% if homes are well designed and in keeping with the local area, a statistic that is backed up by research reports on this issue.[3] It was concluded by the Department for Communities and Local Government (DLCG), therefore, that higher rates of community involvement in the planning and development process could lead to greater acceptance rates for new development.

Box 3.1 Compulsory pre-application consultations between prospective developers and local communities

The government's intention behind compulsory pre-application consultation was set out in an impact assessment[4] prior to the drafting of the legislation:

> The new requirement for developers to undertake compulsory community engagement prior to the submission of planning applications will help deliver the principles of Localism and increase the amount of high-quality planning permissions delivered through the planning system by:

- Promoting the involvement of local communities in the development of significant proposals that will affect them.
- Reducing the number of objections to major planning applications after they have been submitted.
- Promoting better-quality planning applications.
- Providing an opportunity for parties to achieve early consensus on controversial issues before proposals are finalised.
- Providing an inclusive and transparent approach to the consideration of planning applications.
- Complementing other planning reforms aimed at empowering communities, particularly neighbourhood planning.

The policy objectives are to (i) increase community engagement in the planning system and allow communities the opportunity to shape their neighbourhoods, (ii) reduce the costs of the planning process and speed up the system, and (iii) increase the number of high-quality, major applications agreed.

It was the intention of the bill that, while previously consultation was carried out through either a developer's genuine desire to consult or because it was felt necessary to gain the support of the planning committee, a legal requirement to consult would be enshrined in law. In reality, the Localism Act omitted the requirement to consult on planning applications in England and Scotland except in planning applications for wind turbines. This is set out in the Development Management Procedure Order 2015 Part 2 Section 3.

In Wales, however, the requirement to consult on major developments became mandatory on 1 August 2016. And in England and Scotland, although there is no legal stipulation, there is an expectation within most local authorities that consultation will be carried out on all medium- and large-scale planning applications, and there is much anecdotal evidence to suggest that the more comprehensive the consultation, the more quickly the planning application will progress.

The community infrastructure levy

The community infrastructure levy (CIL) was also reformed through the Localism Act. CIL is effectively a development tax, intended to counterbalance the impact of development in a specific community. It was introduced in the Planning Act 2008 as a replacement for Section 106 payments and as a tool for local authorities to deliver community infrastructure. Following the Community Infrastructure Levy Regulations (2010), new development which creates net additional floor space of 100 square metres or more, or creates a new dwelling, is potentially liable for the levy.

Prior to the introduction of CIL, Section 106 agreements had been negotiated between the local authority and the developer on a case-by-case basis. On many occasions the developer would have a specific cause in mind – perhaps a new playground or the creation of a nature reserve. Invariably, this not only benefitted the new development, but presented an opportunity for the developer to create and foster good community relations, as well as gaining some positive PR locally. The introduction of CIL removed this advantage, as CIL payments were received directly by the local authority and spent anonymously.

However, in April 2013 the government brought into force legislation made under the Localism Act whereby if a neighbourhood plan exists in a community which is subject to new development, 25% of the money raised from new development is to be made available to the organisation responsible for the neighbourhood plan (usually a neighbourhood forum or parish council). In many cases, therefore, there is now a clear link between a new development and the funding received by the neighbourhood group.

The abolition of regional development agencies and the reintroduction of the local plan

Under the banner of Localism, regional development agencies were abolished in 2012. With a starting budget of £1.4 billion, economic development was transferred to local enterprise partnerships (LEPs) which cover a smaller geographic area than that of their predecessors.

Strategic planning powers were simultaneously transferred to local authorities in the form of local plans. Today's local plans, which set out local planning policies and identify how land is to be used, carry considerably more weight than the local development frameworks that had previously been produced by local authorities.

All local authorities are required to have an approved local plan in place by early 2017. By May 2016 only 70% had done so, and under powers set out in the Housing and Planning Act 2016, where local authorities fail to do so, the government will intervene to arrange for the plan to be written, in consultation with local communities.

A duty to cooperate

While strategic planning now takes place on a local level, there is a requirement for local authorities to discuss major schemes which cross administrative boundaries through the duty to cooperate. The government's guide to Localism[5] states:

> In many cases there are very strong reasons for neighbouring local authorities, or groups of authorities, to work together on planning issues in the interests of all their local residents. This might include working together on environmental issues (like flooding), public transport networks (such as trams), or major new retail parks.
>
> In the past, regional strategies formed an unaccountable bureaucratic layer on top of local government. Instead, the Government thinks that local authorities and other public bodies should work together on planning issues in ways that reflect genuine shared interests and opportunities to make common cause. The duty requires local authorities and other public bodies to work together on planning issues.

The National Planning Policy Framework (2012)

Just a year after Localism shook up the planning system, the Coalition government introduced the National Planning Policy Framework (NPPF) in 2012, with largely identical intentions: to create more homes through clarifying and speeding up the planning system. The NPPF replaced the long and unwieldy Planning Policy Statements and consolidated more than 1,000 pages of policy across 40 documents into a single document of just 50 pages.

Perhaps the most significant aspect of the NPPF as concerns consultation is the 'presumption in favour of sustainable development': the premise that proposed developments, if sustainable, should receive planning consent.

In theory, this puts all potential developers on a very positive footing as they consult with local residents, but equally, entrenched attitudes towards development cannot be reversed simply by a change in government policy.

Box 3.2 The NPPF[6] on the presumption in favour of sustainable development

At the heart of the National Planning Policy Framework is a presumption in favour of sustainable development, which should be seen as a golden thread running through both plan making and decision-making.

For plan making this means that:

- Local planning authorities should positively seek opportunities to meet the development needs of their area;
- Local plans should meet objectively assessed needs, with sufficient flexibility to adapt to rapid change, unless:
  - Any adverse impacts of doing so would significantly and demonstrably outweigh the benefits, when assessed against the policies in this framework taken as a whole; or
  - Specific policies in this framework indicate development should be restricted.

For decision-making this means that:

- Approving development proposals that accord with the development plan without delay; and where the development plan is absent or silent or relevant policies are out of date, granting permission unless:
  - Any adverse impacts of doing so would significantly and demonstrably outweigh the benefits, when assessed against the policies in this framework taken as a whole; or
  - Specific policies in this framework indicate development should be restricted.

Furthermore, the NPPF strengthens the importance, already present in planning law following the previous year's Localism Act, of pre-application involvement and consultation.

Box 3.3 The NPPF[7] on pre-application engagement and front loading

- Early engagement has significant potential to improve the efficiency and effectiveness of the planning application system for all parties. Good-quality pre-application discussion enables better coordination between public and private resources and improved outcomes for the community.

- Local planning authorities have a key role to play in encouraging other parties to take maximum advantage of the pre-application stage. They cannot require that a developer engages with them before submitting a planning application, but they should encourage take-up of any pre-application services they do offer. They should also, where they think this would be beneficial, encourage any applicants who are not already required to do so by law to engage with the local community before submitting their applications.

- The more issues that can be resolved at pre-application stage, the greater the benefits. For their role in the planning system to be effective and positive, statutory planning consultees will need to take the same early, proactive approach, and provide advice in a timely manner throughout the development process. This assists local planning authorities in issuing timely decisions, helping to ensure that applicants do not experience unnecessary delays and costs.

- The participation of other consenting bodies in pre-application discussions should enable early consideration of all the fundamental issues relating to whether a particular development will be acceptable in principle, even where other consents relating to how a development is built or operated are needed at a later stage. Wherever possible, parallel processing of other consents should be encouraged to help speed up the process and resolve any issues as early as possible.

- The right information is crucial to good decision-taking, particularly where formal assessments are required (such as environmental impact assessment, habitats regulations assessment and flood risk assessment). To avoid delay, applicants should discuss what information is needed with the local planning authority and expert bodies as early as possible.

- Local planning authorities should publish a list of their information requirements for applications, which should be proportionate to the nature and scale of development proposals and reviewed on a frequent basis. Local planning authorities should only request supporting information that is relevant, necessary and material to the application in question.

- Local planning authorities should consult the appropriate bodies when planning, or determining applications, for development around major hazards.

- Applicants and local planning authorities should consider the potential of entering into planning performance agreements, where this might achieve a faster and more effective application process.

Box 3.4 The NPPF[8] on consultation by local planning authorities on individual planning applications

After a local planning authority has received a planning application, it will undertake a period of consultation where views on the proposed development can be expressed. The formal consultation period will normally last for 21 days, and the local planning authority will identify and consult a number of different groups.

The main types of local planning authority consultation are:

- Public consultation – including consultation with neighbouring residents and community groups.
- Statutory consultees – where there is a requirement set out in law to consult a specific body, who are then under a duty to respond providing advice on the proposal in question.
- Any consultation required by a direction – where there are further, locally specific, statutory consultation requirements as set out in a consultation direction.
- Non-statutory consultees where there are planning policy reasons to engage other consultees who – whilst not designated in law – are likely to have an interest in a proposed development.

Following the initial period of consultation, it may be that further additional consultation on changes submitted by an applicant, prior to any decision being made, is considered necessary.

Finally, once consultation has concluded, the local planning authority will consider the representations made by consultees, and proceed to decide the application.

In its response to the Select Committee Inquiry into the Operation of the National Planning Policy Framework in 2015,[9] the government reported a successful outcome of the NPPF, as follows:

- Plan making has significantly improved: 80% of local planning authorities have at least published their plan so are at an advanced stage, and 62% of local planning authorities now have an adopted local plan in place (compared to 17% in 2010).
- Planning permission was granted for 240,000 new homes in England in the year to September 2014.
- Two-thirds of appeal decisions are in line with the councils' original determination – 99% of decisions are right first time with only 1% of applications overturned on appeal.
- Over 1,200 communities across England are now involved in bringing forward Neighbourhood Plans to shape what gets built where in their

local area. This means that more than six million people are living in areas undertaking Neighbourhood Planning.

- The amount of green belt land remains constant, covering over 13% of England.

Devolution deals (2013)

Devolution deals were put forward in 2013 as a means of continuing the theme of Localism by devolving power from central government to local areas in England. Devolution deals allow consortiums of local authorities to take responsibility for economic development and public services, including spatial planning. The first devolution deal was announced by the government and the Greater Manchester Combined Authority in November 2014 and by April 2016, ten devolution deals had been agreed. Seven of the ten combined authorities had taken on responsibility for preparing spatial planning frameworks. Furthermore, it is believed that some city-regions will also set up development corporations and take responsibility for calling in planning applications.

The Growth and Infrastructure Act (2013)

During 2013, the Coalition government introduced two further pieces of legislation which were designed to speed up the planning process, in both cases doing so by exerting powers over local planning authorities. The first was the Growth and Infrastructure Act which was put in place to simplify and speed up the planning process for large-scale planning applications, through measures which included:

- Allowing developers to make certain planning applications directly to the secretary of state rather than to a 'proscribed' local authority. An authority could be proscribed if they failed to determine major planning applications within 13 weeks or if 20% of major applications had previously been overturned on appeal.
- Providing a 'fast-track' route for large-scale applications. This is described in detail in Chapter 13.

While the 2013 Act and the 2015 Infrastructure Act which followed it provide the opportunity for schemes to be given consent in situations where local authority failings make this otherwise impossible, local consultation is no longer overseen at a local level: instead it is either part of the NSIP process or part of the standard planning application process but determined by central government. Proposals which are determined at a national level can face problems with community relations following planning consent due to a perceived lack of local involvement in the planning decision.

Change of use consent (2013)

A second and very controversial initiative of a similar nature was the change to permitted development rights which were brought about in May 2013 initially on a temporary basis following an amendment to The Town and Country Planning (General Permitted Development) (England) Order 2013. The amendment allowed the change in use of buildings from B1(a) (offices) to C3 (dwelling houses), subject to a prior approval process by the local planning authority and, crucially, without planning consent. This meant that many office buildings were converted to residential use devoid of any public consultation. In the following years, change of use was allowed on a range of other use types, some of which were made permanent.

Housing and Planning Act (2016)

Publishing its proposed legislation in 2015, the government said the Housing and Planning Bill would kick-start a 'national crusade to get 1 million homes built by 2020' and transform 'generation rent into generation buy'.[10]

Enacted in May 2016, the legislation certainly had a substantial impact on the market. Its many measures included an extension of the right-to-buy for housing association tenants, the promotion of self-build and custom-built housebuilding, changes in relation to council rent levels and tenancies, and the instruction that starter homes be delivered on all 'reasonably-sized' sites. The Act also clarified that brownfield land registers (to provide housebuilders with up-to-date information about brownfield sites available for housing) would, in time, become mandatory for all English councils and included an extension of permitted development rights in relation to change of use.

Changes were also introduced to allow NSIP planning procedures[11] to apply to large-scale housing schemes – something that had been long anticipated and are seen by many as the only efficient means of bringing about large-scale schemes, such as garden cities, which cross administrative boundaries and involve substantial infrastructure.

However, local input in development decisions, particularly at a strategic planning level, is potentially much reduced by the new legislation. First, the Act gives new powers to the secretary of state to take over the neighbourhood plan-making role of the local planning authority: where a neighbourhood area application meets prescribed criteria but has not been determined within a prescribed period, the secretary of state may require a local planning authority to designate all of the entire area. Furthermore, the secretary of state may prescribe the time periods within which local authorities must undertake key neighbourhood planning functions. At the same time, neighbourhood plans are given greater significance in the planning system through a new requirement that a planning application makes direct reference to the way in which the neighbourhood plan was taken into account in drawing up the proposals.

Second, the Act reduces the powers of local authorities in strategic planning. If a local authority has failed in its preparation of a local plan, the Act enables the secretary of state to direct an inspector to suspend the local plan examination, to consider specified matters, to hear from specified parties, or to take other specified procedural steps. The secretary of state may also invite the mayor of London or a combined authority to prepare a local plan for a London borough or a constituent authority of the combined authority, and either approve the document or direct the local authority to consider adopting it – even recovering the costs of doing so. The legislation states that the secretary of state would do so in consultation with local people, though the extent to which this would occur and the effectiveness of a Whitehall-led consultation remains to be seen.

Perhaps an even more significant development is that of permission in principle (PiP): a new, automatic consent for schemes subject to further technical details being agreed by authorities.

The government has identified two types of PiP: 'Allocation PiP' for land allocated for housing-led development in 'qualifying documents' such as development plan documents, neighbourhood development plans and in certain registers, including the brownfield register; and 'Application PiP' for small sites under 10 units where a developer has made an application to the local authority.

With the impetus being to use such measures to boost housebuilding, the government has made it clear that this measure will be used primarily in the case of housing. While it has ruled out this measure being applied to planning applications for fracking, however, it does not rule out other development types – and in circumstances where the site in question is a large-scale mixed-use development, it would be impossible to separate retail, education and leisure use from that of housing.

The government's intention is that in order to achieve technical consent, applicants provide substantial details about the finer details and to consult with communities and other interested parties at this stage. There is a danger, however, that communities confronted with a planning application which already has 'permission in principle' would be unlikely to concern themselves with a consultation on the technical matters, on the basis that the proposals are already approved.

Conclusion

The over-riding ambition of the governments of 2010 and 2015, as far as planning is concerned, has been to deliver an increased number of houses, at increased speed. Localism, in its various guises, was initially seen as the most appropriate means of doing so and introduced some bold and creative tactics. But Localism largely failed to inspire communities to accept development. Consequently, the point at which the Coalition government was replaced with a Conservative government in 2015 saw a likely return to a more centralised planning system, with various threats to take control of local planning were enshrined in law. And while devolution deals are intended to encourage local involvement in planning, the government has also sought to regain powers where it is deemed necessary to meet its ambitious housing targets.

It is perhaps not surprising that, having had expectations raised with the excitement surrounding the Big Society in 2010, many local people are more disillusioned than ever before in their potential control over their neighbourhoods. Compounded with a more acute need for housing, this has resulted in people expressing their concerns more vociferously than ever before.

Notes

1 DCLG (Department for Communities and Local Government) (2011) *A Plain English Guide to the Localism Act.* London: DCLG.
2 Cabinet Office (2011) *Open Public Services* White Paper. London: The Cabinet Office.
3 National Housing and Planning Advice Unit (2010) *Public Attitudes to Housing [Online].* Available www.communities.gov.uk/documents/507390/nhpau/pdf/16127041.pdf [Accessed 12 October 2016]
4 DCLG (Department for Communities and Local Government) (2011) Localism Bill: Compulsory Pre-Application Consultations Between Prospective Developers and Local Communities. London: DCLG.
5 DCLG (Department for Communities and Local Government) (2011) *A Plain English Guide to the Localism Act.* London: DCLG.
6 DCLG (Department for Communities and Local Government) (2014) *The National Planning Policy Framework.* London: DCLG.
7 Ibid.
8 Ibid.
9 DCLG (Department for Communities and Local Government) (2015) Government Response to the CLG Select Committee Inquiry Into the Operation of the National Planning Policy Framework. London: DCLG.
10 Wilson, W., and Smith, L. (2015). *Housing and Planning Bill* [Bill 75 of 2015–16]. Briefing Paper Number 07331, House of Common Library. London: HMSO.
11 See Chapter 13 for a description of the planning process in relation to NSIPs.

4 Societal change and community involvement

Introduction

Since the start of the twenty-first century, social changes have impacted extensively and considerably on community involvement. The purpose of this chapter is to enable those involved to better understand the changes. A necessary starting point is to consider what is meant by 'community' and whether 'community' has any real meaning in the twenty-first century. We will also address changing levels of involvement in planning and determine how, or whether, people wish to be consulted – enabling planners, developers and local authorities to create effective strategies for involvement. The means by which people choose to communicate will be considered, though the most considerable change of all – the impact of online communication – will be discussed in depth in the following chapter.

The *Oxford English Dictionary* describes 'community' as 'a body of people living in one place, district or country', and 'a body of people having religion, ethnic origin, profession etc. in common'.

In planning, we tend to regard community in geographical terms, which is perhaps inevitable as a new development has a physical impact on a specific location. Furthermore, a local authority's consultation will generally be aimed at the residents of that specific geographic jurisdiction. The *OED*'s second definition, however, should not be overlooked, as we will discuss later.

Doak and Parker (2012)[1] offer a thought-provoking consideration of the term:

'Community is a well-worn term that has been used and misused in public discourse and broadly across the political and social sciences. In planning terms much of the activity of planners is justified as being in the public interest but more and more the notion of community is attached to variety of planning processes, policies and actions. This common association of planning activity to and for community as both an end and a stakeholder group justifies an exploration of the term and its relevance for planners and in planning practice.

'Community was seen as a political ideal in the ancient world, where citizens could participate in public affairs as part of the community. The concept has developed such that "community as belonging" has come to be viewed both as a past state and as a desirable aspiration. Hobsbawm[2] pointedly observes that, "Never was the term community used more indiscriminately and emptily than in the decade when communities in the sociological sense became hard to find in real life." Even more pessimistically, Bauman[3] indicates that predilections towards recovering or developing community ignore the likelihood that it never existed in the first place.'

Doak and Parker's sentiment eloquently voices the widely held view that the term 'community' can all too often be used nostalgically, euphemistically and even patronisingly, and that in some cases it is an artificial concept. To an extent, this is true in community involvement: there is also a tendency for those running consultations to use 'community' as a convenient catch-all for the streets neighbouring the proposed new development or within a one-, five- or ten-mile radius from the site. There are clear practical reasons for doing so, but it would be quite wrong to assume that a line drawn on a map by someone with little knowledge of the neighbourhood constitutes a community. Furthermore, it is equally inaccurate to assume that the supposed 'community' is a single body and likely to respond with a single view: as we know from our own neighbourhoods, rarely does everyone on one street, let alone the wider neighbourhood, have an identical view on any one matter.

Changing geographic communities

Diversification is, in fact, one of the most significant changes in the concept of community.

The recent hey-day of the community was perhaps at a time between the end of the Second World War and the fragmentation which took place during the 1980s. Picture a scene from the 1950s or 60s: a street of terraced houses in a city previously damaged by war but still united by the blitz spirit, where the majority of occupants came from a similar social background, spoke regularly over the garden wall, read the local newspaper and attended the same local schools, social clubs and churches. They literally sang from the same hymn sheet: experiences were shared and there was an element of mutual trust, understanding and support.

Half a century later, it is extremely unlikely that the same residents, or indeed their families, still reside on that street. Increased multiculturalism will have led to greater diversity. Increased property prices and greater fluidity in the property market will have resulted in some houses having been extended, with others converted into flats, leading to a wider demographic. Attendance at the churches has probably declined, while initiatives to allow greater parental choice will mean that not all of the children attend the local school. Few residents will work within walking distance of their homes, with many commuting to the city centre or to a different town or city entirely. Furthermore, the community support officers who were employed to establish community relations post-war and the community arts projects which were popular during the 1960s and 70s are no more, as the case with many local newspapers.

The twentieth-century community was by no means a utopia, but from a community involvement point of view, a geographically defined community was certainly a convenient starting point. There is significantly less homogeneity in local communities today. Global communication increasingly takes the place of local communication – whether in politics, business or leisure time. Increased car ownership, the availability of cheap flights and the ease with which travel plans can be made online has vastly increased the size of the communities within which people operate. Today it is easier to send an email to someone on a different continent than to visit a next-door neighbour.

Changing communities of interest

It follows that the dissipation, and perhaps decline, of geographic communities results in the rise in communities of interest. This is perhaps best illustrated in the context of our leisure time. Previously, individuals' experience of music would have been through participation, or attending live music in concert halls, pubs and social clubs. Today, much of the music that we listen to is online or through electronic means. Live music is still popular today, of course, but it is frequently consumed from across the Atlantic via the internet. Social media has enabled people to take part in live discussions in relation to a band or performance, and increasingly the internet provides opportunities for collaboration online. In sports, too, participation and support of local clubs has declined partly due to the wide availability of sports coverage from across the world, while fan clubs, Facebook groups and to-the-minute discussions on Twitter are increasing levels of interaction irrespective of geographical boundaries.

It follows that in development, the community of interest is potentially global. Bicester Village, a designer outlet centre in Oxfordshire, attracts 14,000 Chinese visitors each year.[4] Even for a single, specialist shop, the community of interest may be worldwide. The same is true of opposition to a development proposal: the community of interest, where a development involves building on open countryside, the demolition of a building of historical interest or the destruction of an important natural habitat, will be considerable and may come from across the country or perhaps the world.

So communities certainly exist in the twenty-first century, but on a very different basis to those that went before them: communities are more likely to be linked by interest than by geography than they were previously, and membership may be more passive, virtual and transient.

Of course, planning is usually with reference to a geographical feature, and the immediate neighbours will remain a priority. But developers should also invest time in understanding the communities of interest that may put forward their point of view, whether in supporting a planning application or opposing it.

The rise in single-issue and direct action groups

Communities of interest are perhaps at their strongest and most eloquent when they gather collate to form a single-issue or direct action group. Traditional forms of civic involvement have declined during the twenty-first century: trades union membership has almost halved since the late 1970s and now comprises less than a quarter of the workforce.[5] Yet membership of special interest groups has increased substantially. Nearly 4.5 million people, or one in 10 UK adults, is now a member or supporter of one of Britain's environment and conservation groups.[6]

Single-issue groups are those which exist to lobby on a specific subject. As such, they tend to be motivated by a notion of injustice or threat, or a need to bring about change. Successful single-issue groups, such as Make Poverty History, the Taxpayers Alliance or Fathers for Justice, were formed with a single imperative that unites members and as such they promote their messages very effectively, whether through protests, stunts or the media. There are many single-issue groups which impact on planning, from international organisations, such as Greenpeace, to local conservation groups. Because such groups have been founded on the basis of a specific cause, they can provide substantial opposition to a new scheme. The campaigning power of the internet means that despite a small budget, even a small membership can have a considerable impact. And because national groups quickly reorganise on a local level in relation to a specific proposal, they can have local relevance while drawing on their national strength.

Single-issue groups do play an important role in the planning process. Where they have shown an interest in a proposal, every opportunity should be made to engage with them, to understand their point of view, to correct any misapprehensions which may exist and to take on board all feedback which is relevant to the planning application. The consequences of failing to engage with powerful interest groups will be significantly outweighed by time taken to consult with them. And single-issue groups are not necessarily a negative force in planning: developers frequently find that where a neighbourhood has several groups in place in response to an unpopular former planning application, those very groups may lend their support to a new proposal.

Special interest groups can also be extremely constructive in the case of a specialist facility. A developer of a specialist sports centre, for example, would benefit from consulting with those who already enjoy the specific sport. Not only will those with an interest provide valuable feedback to a consultation, but they may be extremely helpful in promoting and supporting it at a later stage in its development.

Increasingly, largely in response to the campaigning power of the internet, there has been an increase in the number of direct action groups which exist simply to campaign, rather than having formed around a specific issue.

Largely internet-based, groups such as these have a strong campaigning capability and considerable power to draw attention to an issue, locally, nationally and internationally. But most petitions simply state their support/opposition, and as this is typically the extent of their involvement, it can be difficult to form any meaningful dialogue. Therefore the challenge is to identify, where possible, those behind the campaign and having done so, seek some meaningful engagement. Equally important is the need to mitigate any negative publicity, both online and offline, correcting misapprehensions and providing reassurance where necessary while also putting in place additional measures to promote more positive messages.

Box 4.1 38 Degrees

38 Degrees is one of the UK's biggest campaigning communities. It is accessible via a website which allows individuals to start a petition, organise meetings and generate support.

Campaigns frequently extend beyond the web, often involving lobbying tactics such as phoning MPs and arranging newspaper advertisements. Local groups can be formed, meetings arranged and celebrations often take place when campaigns are won.

The organisation, which has a very small staff base in relation to the size of its community, is funded by small donations from its members.

38 Degrees has been influential in planning issues, such as collecting 68,769 signatories in opposition to fracking, 1,939 to prevent the sale of school fields, and 19,617 to urge the Scottish government to hold a public inquiry in relation to Donald Trump's plans to create a golf resort on the Scottish coast.

Box 4.2 Complaints Choir

The Complaints Choir began in Birmingham in 2005, when the founders invited people to send complaints and to join workshop meetings. With the help of a local musician, they turned the complaints into a song for the choir and performed it in public.

After its initial success, the idea spread through YouTube, resulting in Complaints Choirs being set up across the world. The organisation now has a website which offers advice to anyone wishing to set up a Complaints Choir.

According to the founder, Tellervo Kalleinen[7]:

> For Complaints Choir, the aim is to invite people to look more closely at the phenomenon of people who spend time complaining. Complaining is a huge source of energy but the question then arises of what do you do with all this energy? If a change happens it will usually start with the activity of complaining about a situation.
>
> So we start by raising general awareness of our tendency to complain and then move on to explore how you can do something with these complaints, even if it's just the experience of making a choir song. We take this energy of complaining and then work to do something uplifting and energising with it, rather than simply dwell on the melancholy and depressing aspects of this.
>
> There can be a big power when people get together if you can play with individual complaints in this way. So basically what we do is about providing this experience. We offer this experience at an individual level but it also works quite well when people perform or experience the songs in public too, whether an audience see the songs in an exhibition, or via a video, or experiences a live performance.
>
> The more the media is bringing us into an information society and raising awareness of the many problems there are all around the world, the more it's becoming a challenge to negotiate who we are within it. I believe giving focus to the things nearby is a very important part of our identity and it is okay to be irritated by so-called small things.

Changing levels of involvement in planning

The first decade of the twenty-first century was a new era in which individuals were given the opportunity to become more active, with significantly more control, whether as a purchaser, a customer or a commentator. This was largely due to increasingly interactive internet-based communications but also brought about through growing customer choice (for example in health and education). The advent of corporate social responsibility (CSR), which first became enshrined in law from the late 1990s, the Freedom of Information Act 2000 and Local Government (Access to Information) Act 1995 formally accepted the increased importance of dialogue with consumers and responsibility to both local communities and society at large. The result was a shift in the balance of power between consumers and companies or organisations.

But the multiple means by which to speak out and the increased likelihood of being listened to does not necessarily translate into an increase in those contributing to discussions on matters such as planning. On the contrary, our busy lives now present so many opportunities to submit a review following a purchase, comment on a Facebook group or complete a council survey that we can be inundated with requests and suffer what has become known as 'consultation fatigue'. Consultation fatigue can also be caused by a local authority running, or allowing, multiple consultations in a specific location, or seeking too much involvement on complex, nebulous planning questions such as in the formation of a local plan. When overwhelmed by multiple requests for feedback, most individuals choose to respond to an organisation with which they have a particular affinity, or an issue that they feel particularly strongly about. It follows, therefore, that those people who have little affiliation to a company proposing a new development in their area and no particularly strong views as to whether it goes ahead or not are less likely to engage with the developer than they might once have been. This group is sometimes, perhaps optimistically, referred to as the 'silent majority'.

Individuals are also reluctant to become involved in planning discussions if they feel that their contribution is unlikely to change anything, perhaps because they have previously taken part in consultations on planning issues and feel that this has had little impact. Overcoming such perceptions requires the developer to select the topics to be discussed very carefully, to commit to genuine two-way consultation, to communicate in a manner that is clear and easy to comprehend (not only in terms of the vocabulary used but also the structure of the consultation), to select consultation tactics which motivate people, and to undertake to report on all consultation responses. Each of these principles will be addressed in detail in Part III of this book.

Box 4.3 Future community research
Comment by Rachel Lopata, director, Community Research

Research and engagement, as we know it, faces a number of challenges over coming years. Big changes are being created by a combination of the impact of 'big data', the rising use of passive monitoring, DIY research tools and waning respondent attention spans. This is a brief look into our crystal ball to try to find out what all this might mean over the coming year and beyond.

There has been an explosion of data – according to a recent estimate from IBM,[8] we've generated 90% of all data in the entire history of the world in just the past two years. 2.5 quintillion bytes of data per day! Traditional researchers are used to dealing with small datasets that they control. In the future, data will come from technology companies, internal company systems and the 'Internet of Things' (the network of physical objects or 'things' embedded with electronics, software, sensors and network connectivity which enables these objects to collect and exchange data.) Survey data will be just one amongst a plethora of data sources. And even this survey data may be collected by clients directly given the availability of easy-to-use computer-assisted research tools. Furthermore, we have become accustomed to the fact that behaviours are often more effectively captured and understood by passive monitoring (e.g. observation or ethnography) rather than asking people questions about their behaviours. 'Nudge' theory is becoming more central to all of our thinking.

Added to this, there seems to be a consensus that respondent attention spans are continuing to fall and that ever more creative approaches are needed to cut through and to engage. Some suggest that micro surveys and intercepts will, in time, replace longer surveys and tracking questionnaires.

Consultation in the future will, therefore, need to be able to understand the wide range of data that comes from various sources and help to make sense of it. It will be necessary to be able to apply a wider range of research techniques – such as ethnography and social media analysis, in addition to traditional surveys and group discussions.

Those carrying out consultations will need to be aware of the shift to more 'agile' research, perhaps focusing on short studies and experiments, frequent feedback, and the ability to react to changing conditions. We see this already in terms of growing interest in co-production and co-design techniques as well as the rising use of online panels and communities.

A sense of place and community relations

The most successful consultations are those which value contributions and use them to benefit development proposals. The most effective way to involve local people and gain constructive responses is to value their understanding. A 'sense of place' can only be acquired over time, through day-to-day experiences of a neighbourhood and an awareness of its history. There are all too many examples of outsiders having simply failed to grasp the symbolic and physical significance of sites with serious consequences for a development. To avoid accusations of a 'one size fits all' or 'prototype' scheme, developers should work, both with the local authority and local residents, to acquire local knowledge.

Clifford (2013),[9] in a study of the reaction of frontline planning officers to participation by the public, found that having a local sense of place is useful and participation can help to explore its significance. One planning officer responded:

> I think that it [participation] actually improves the product, quite often people will know things, will have ideas that you as a planner don't have, either because you can't literally know everything particularly if you are dealing with a different area, and different people come at things from a different perspective so that what you end up with is a richer response.

Political initiatives, such as the removal of regional planning strategies and the introduction of Localism, have been put in place to increase local involvement in planning at the expense of more regional control and build on the social capital within a community. With the support of neighbourhood plans averaging at 87%,[10] it would seem both that residents welcome locally generated plans and that this has potential to reduce consultation fatigue.

Conclusion

To run a successful involvement programme, developers, local authorities and planning consultants must be aware of changing structures in society, changing means of communication and changing attitudes towards community involvement. Significant developments have already taken place in the twenty-first century and the pace of change is not slowing: it will be important to keep up to speed with changes which both impact on our ability to consult effectively and also present opportunities for new ways of doing so.

'Generation Z' is a sharing generation: the post-millennials have grown up with Facebook, Survey Monkey and 38 Degrees, and there is a strong feeling among the development industry that as these individuals reach adulthood, they will be easier to involve than their parents were. Inspiring them to become, and remain, involved in planning requires the development industry to keep abreast of change and use it positively. The next chapter addresses the role of technology in this process.

Notes

1 Parker, G., and Doak, J. (2012) *Key Concepts in Planning.* Los Angeles: Sage.
2 Hobsbawm, E. (1994) *The Age of Extremes.* New York: Pantheon Books.
3 Bauman, Z. (2001) *Community.* Cambridge: Polity.
4 *Chinese Tourists Visiting Oxfordshire 'Double in Decade'* (2015) [Online]. Available www.bbc.co.uk/news/uk-england-oxfordshire-33252945
5 Trades union membership fell from 13,212 in 1979 to 7,011 in 2013–2014: GOV. UK. (2016) *Trade Union Statistics 2015 – Publications – GOV.UK* [Online]. Available

www.gov.uk/government/statistics/trade-union-statistics-2015 [Accessed 19 September 2016]

6 *Passionate Collaboration? Taking The Pulse Of The UK Environmental Sector* [Online]. Available www.greenfunders.org/wp-content/uploads/Passionate-Collaboration-Full-Report.pdf [Accessed 19 September 2016]

7 Stephansen, H. (2016) *Complaints Choir: What Is It?* [Online]. Available www.open.edu/openlearn/people-politics-law/politics-policy-people/participation-now/resources/complaints-choir-what-it [Accessed 19 September 2016]

8 *IBM – What Is Big Data?* (2016) [Online]. Available www-01.ibm.com/software/data/bigdata/what-is-big-data.html [Accessed 19 September 2016]

9 Clifford, B. (2013) Rendering Reform: Local Authority Planners and Perceptions of Public Participation in Great Britain. *Local Environment,* 18(1), 110–131.

10 Geoghegan, J. (2016) Close Call. *Planning,* 2036, 15.

5   The impact of the internet on community involvement

Technological changes have, without doubt, had the greatest impact on community involvement so far this century: the way in which we access information, our ability to debate and discuss online and initiatives by both local and national government to transfer services online have created a position whereby no planning application is without an internet presence.

That is not to say that all developers and planning consultants actively engage in online involvement but that, whether the development team intends it or not, proposals for a substantial development are very likely to be the subject of lively discussions on websites, blogs and social media.

This chapter will therefore address the external technological change impacting on planning and conclude with a brief section on the significant opportunities that online engagement offers – something that will be explored in greater detail in Chapter 16.

Growing internet usage

It is widely understood that use of the internet has increased year on year, and will continue to do so until complete saturation occurs. In some areas and in some age groups, this point has almost been reached. In Great Britain in 2016, the internet was used daily by 82% of adults – a 47% increase over 10 years. No fewer than 99.2% of adults aged 16 to 24 years were recent internet users – though this contrasted sharply with 38.7% of adults aged 75 years and over.[1] And although, with just over one-third using the internet, the older population has some way to go before it reaches saturation point, women aged 75 and over are the fastest growing group, up 169.0% from 2011.[2]

The way in which we use the internet also continues to grow and diversify (Figures 5.1–5.4). The year 2016 saw 70% of adults access the internet 'on the go' (away from home or work) using a mobile phone or smartphone – double the proportion five years previously;[3] 77% of adults bought goods or services online,[4] and 61% of adults used social media.[5]

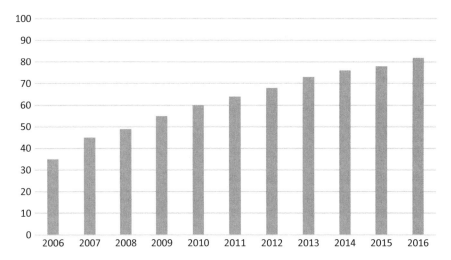

Figure 5.1 Percentage of daily internet usage by adults, 2016

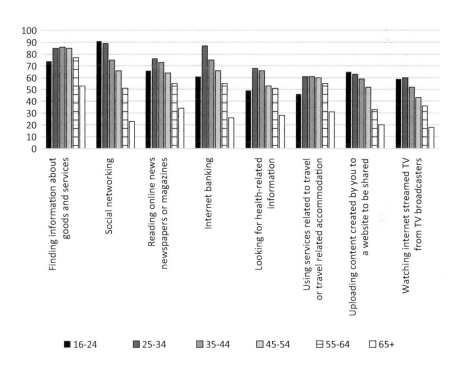

Figure 5.2 Percentage of internet activities by age group, 2016

Changing forms of communication

The presence of the internet does not only provide a new platform on which to communicate: it changes the manner in which we communicate. The world of online communication is, by and large, very democratic and non-hierarchical. Every user has the potential to broadcast a message to millions worldwide at the touch of a button, and consequently the concept of

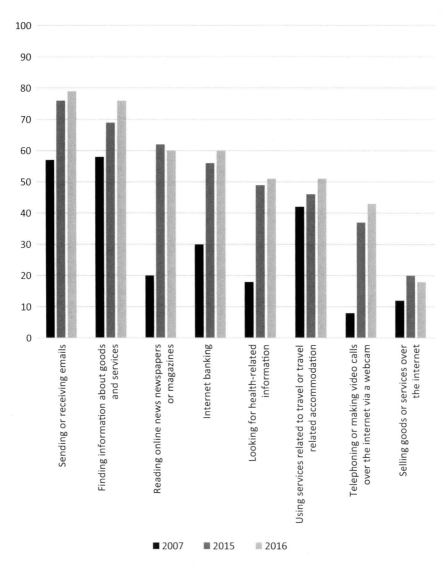

Figure 5.3 Percentage of internet activities carried out by UK adults, 2016

'citizen journalism' is growing by the day. That message will then appear unaltered and without confusion of an external influence. In principle if not in practice, a level playing field has been created, and large organisations which were once able to use their position to influence are now exposed to previously unencountered levels of challenge and opposition online. Consumers now have greater expectations from organisations and the power to ask for information publicly. Through the internet we have a greater opportunity to be informed and also a greater capacity to seek knowledge. But where the lack of an organisational filter removes the need for checks to be made, misinformation can occur.

With the advent of Web 2.0 in 2004, the extent to which people could collaborate, comment and share information increased, resulting in the internet ceasing to be only a vehicle through which information could be sought to an opportunity to both broadcast information and enter into dialogue on a number of levels. The speed by which information now travels would be inconceivable to someone living in the first half of the twentieth century – through websites, blogs, social media, apps and email, information can be both sought and imparted within seconds. Not only does the initial message occur immediately, but a post, email or Tweet can be shared with similar speed, 'snowballing' and thus reaching millions. Many websites will now enable this to occur automatically – composing a Tweet or a link to Facebook the moment a purchase has been made or a poll completed – crucially, with little or no effort on the part of the author/publisher. In fact, the curating and sharing of a piece of information can occur devoid of human interaction: the algorithms that power Facebook and Google are responsible for much of the content that we consume.

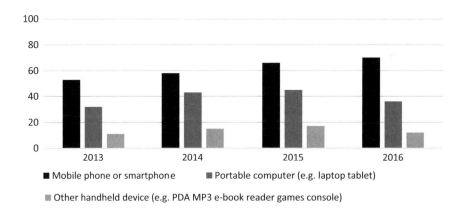

■ Mobile phone or smartphone   ■ Portable computer (e.g. laptop tablet)

▨ Other handheld device (e.g. PDA MP3 e-book reader games console)

Figure 5.4 Percentage of adults aged 16+ in Great Britain accessing the internet away from home or work, 2016

The American University Center for Social Media[6] identified internet usage as falling into five categories: choice, conversation, curation, creation, and collaboration. In a planning context, these behaviours might be described as follows:

- **Choice**: finding information on local plan formation, policies and planning applications though search engines, recommendations (on- or offline), news feeds and niche sites.
- **Conversation**: entering into debates on discussion forums, blogs and microblogs, taking discussions into new forums by sharing links and mobilising action.
- **Curation**: selecting and drawing together information on blogs to form powerful arguments, carefully targeted to specific groups; posting and reposting views and suggestions and sharing links.
- **Creation**: posting brand new multimedia content, including text, images, audio and video rather than simply responding to information posted by a local authority, developer or government body.
- **Collaboration**: creating groups of support or opposition for the purposes of campaigning both online and offline.

As the capabilities of the internet, along with internet usage, grow, the opportunities for involvement within each of these categories will undoubtedly increase, and individuals' behaviour online is likely to become less passive and more powerful. Developers who opt not to have an online presence, or who install a consultation website with no mechanism for dialogue, run the risk of their scheme being debated on closed blogs and Facebook groups and as such will be unaware of any mounting objection until it becomes too late to prevent it. The industry must accept the changing communications landscape, monitor sentiment and proactively encourage constructive consultation online. The means by which developers can do so is explored in Chapter 16.

Use of the internet by national government

Central government aspires to break down the misconception that the digital world is separate from everyday living. The digital transformation of government is led by the Government Digital Service (GDS), part of the Cabinet Office, which works across all government departments. It was set up following a set of recommendations by the then UK Digital Champion Martha Lane Fox in 2010.[7] The GDS aims to provide online services to millions of users and in doing so, deliver £1.7 billion in annual savings though efficiencies such as reduced staff time in processing transactions and reduced accommodation, postage and packaging costs.

The GDS has also put in place the digital service standard, a set of 18 criteria to help government create and run good digital services. These are shown in Box 5.1.

Box 5.1 Digital service standard

1 Understand user needs
2 Do ongoing user research
3 Have a multidisciplinary team
4 Use agile methods
5 Iterate and improve frequently
6 Evaluate tools and systems
7 Understand security and privacy issues
8 Make all new source code open
9 Use open standards and common platforms
10 Test the end-to-end service
11 Make a plan for being offline
12 Make sure users succeed first time
13 Make the user experience consistent with GOV.UK
14 Encourage everyone to use the digital service
15 Collect performance data
16 Identify performance indicators
17 Report performance data on the performance platform
18 Test with the minister

In January 2013, the government gave itself 400 days to transform 25 major services, making them 'digital by default' and simpler, clearer and faster to use. This was known as the Transformation Programme. Crossing eight government departments, it set about redesigning a set of digital exemplars based on the needs of users, rather than that of government. Services which are available online as a result include registering to vote, renewing a patent, student finance, visas, apprenticeships, digital self-assessment, PAYE, vehicle management, personalised registration and redundancy payments.

The programme has met with considerable success. This is perhaps best demonstrated by the online voter registration system, which allows citizens to apply to register to vote in approximately three minutes. In the first two months, 82% of the one million people who registered to vote used the new online service rather than paper-based methods, with around a third of those accessing the service through a smartphone or tablet. The GDS claims there is over 90% satisfaction with the new online service.[8]

Like many other public services, planning is fundamentally affected by initiatives to move services online. The Planning Portal was set up in 2002 to allow planning applications in England and Wales to be processed electronically. It also provides substantial amounts of information, both as briefing notes and interactive guides.

Use of the internet by local authorities

Local government is the body responsible for processing planning applications. And in common with all public-facing organisations, local authorities have understood the need for and the opportunities of online communication. With an average of 100,000 transactions per year, no local authority today is without a website and most have apps and Twitter accounts, while many also have Facebook and Instagram pages. From providing rubbish collection timetables to facilitating school admissions, local authority websites are enormous, diverse and potentially a huge opportunity.

All planning applications, local plans and neighbourhood plans are now available to view online, thus reducing the need for residents to make an appointment to view plans in a local planning department and substantially increasing the potential for documents to be viewed. Responses are generally encouraged via the website, in fact some local authorities no longer accept comment via telephone and postal addresses, for the planning team tend to be well hidden. It is not surprising, therefore, that local authority planning teams encourage developers to use consultation websites when conducting large-scale consultations.

At the same time, local authority staff are increasingly using mobile technologies, route-planning tools and video conferencing to increase efficiency.

The online revolution is not without its challenges, and this is particularly true in the context of reduced funding for local government. To local authorities, the balance between using online communication to introduce greater efficiencies and improve customer service, while running the risk of failure, is a precarious one. Other challenges include the need to substantially rethink services to ensure that online contact is as effective as offline; providing a service which is appropriate to all demographic, age and ability groups; overcoming potential technical failings; and encouraging both the public and the workforce to accept new methods of communication.

Most local authorities would probably agree that the benefits of online communication vastly outweigh the challenges. Increased speed, reduced cost and the potential to increase efficiencies are not the only benefits to online communication. Using the internet also enables local authorities to collate information, to analyse and report on it with greater ease and reduced cost. Services can be communicated and promoted effectively through a variety of online channels, including social media. And the remodelled services have on many occasions resulted in an infinitely more effective product. Furthermore, initiatives such as the Public Services

Network (PSN) provide an infrastructure which allows both greater access to public services online and improved communication between councils and government as a whole.

The internet has also significantly benefited community leadership and facilitated the Localism agenda. The use of websites and social media in neighbourhood planning, for example, has enabled communities to organise themselves and drive forward initiatives in ways which would have been impossible previously. As a result of webcams increasingly being present in planning committee meetings, planning decision-making can now be viewed online, enabling all those involved in the process to gain a greater understanding of the process and the way in which a specific planning application is viewed.

Box 5.2 Case study: consultation website

Newcastle City Council runs a consultation website which enables residents to take part in the consultations run by the council. Let's Talk Newcastle (letstalknewcastle.co.uk) hosts online surveys and forums and also allows users to create their own polls and forums. The website integrates with Twitter, Facebook and YouTube for promotional purposes. Users are alerted when new consultations are added.

For each consultation, background information is provided and individuals can choose from a selection of means by which they may respond. These are identified by icons and include polls, offline events, surveys, focus groups and topic walls (forums).

The campaigning power of the internet

Protesting against the status quo, lobbying politicians, campaigning against organisations, cause-related fundraising and political campaigning have flourished in line with the accessibility of internet communications. Whether to inform, mobilize or bring about direct action, the abundance of communication tactics now available enables anyone to run a powerful online campaign. From local campaigns, such as a group of residents campaigning to save a beloved field from development, to international organisations raising awareness of a specific issue such as fracking, the internet can have a major impact on planning proposals.

As an innovative, informative, interactive, and creative tool, online communications (specifically as a result of Web 2.0) have enabled increasingly sophisticated campaigning tactics which are almost certain to mature as technology advances. At little or no cost, a single individual is able to send a powerful message to likeminded audiences, the traditional media, and ultimately the public at large, with significant consequences.

Box 5.3 Online political campaigns

Barack Obama's 2008 US presidential campaign is now recognised as the first online presidential campaign. By using interactive Web 2.0 tools, Obama's campaign changed the way politicians organise supporters, advertise to voters, defend against attacks and communicate with constituents. It relied heavily on social media (including Facebook, YouTube and a custom-generated social engine) to engage voters, recruit campaign volunteers, raise campaign funds and to reach new target segments. YouTube videos were found to be particularly successful, perhaps because viewers chose to watch them, having sought out the information or received links from a friend. This results in a much more appreciative audience compared to an evening's television viewing being interrupted by an unwelcome party political broadcast. It was estimated that Obama's YouTube broadcasts were watched for 14.5 million hours, and that to buy the equivalent in television advertising would have cost $47 million.[9] Furthermore, Obama's team used the web to raise half a billion dollars from 6.5 million donors, collecting more than 13 million email addresses in the process.[10]

It would vastly understate the power of the internet to describe 'online campaigning' as a single tactic. Most offline tactics can be replicated online and thus the presence of the internet immediately doubles the tools available to campaigners. Box 5.4 demonstrates the likely stages that a typical campaign involves and the online tactics which might be used at each stage.

Box 5.4 The stages and tactics of an online campaign

Research

- Gather relevant facts and statistics in support of a campaign.

Communicating the purpose of the campaign

- Create a website to collate and present data.

Broadcast

- Use Facebook, Twitter, Instagram, YouTube and email to promote the campaign.

Networking and interacting

- Find organisations with similar values, motivations and interests online and collaborate to add strength to the campaign.
- Use internet tools such as MeetUp to arrange meetings and protests.
- Permeate existing campaigns and comment on topical issues.
- Use the technology behind sites such as Google and Facebook to promote message socially through algorithms.

Polling

- Create an online poll using free services such as Survey Monkey or PollEverywhere.com.

Force dialogue

- Ask questions of governments and organisations and demand an answer online. (The absence of an answer is often of more value than an actual answer in this case.)
- Make Freedom of Information requests via email.

Growing the campaign

- Use crowdfunding to raise the necessary capital to invest in the campaign.

So what defines online campaigning as opposed to the offline campaigns of the last century?

- **Research and analysis:** automatic alerts services, website analytics and social media monitoring are just some of the tools available to online campaigning which would take considerable time, effort and expense offline. Furthermore, the power of the web to quickly locate planning applications, local authority planning documents and government or pressure group documents and identify potential supporters considerably benefits campaigners.
- **Ease:** the internet is becoming increasingly mobile, intuitive and accessible and thus significantly more user friendly than offline alternatives. Taking part in a campaign in opposition to planning proposals online can be as simple as receiving a link via email and clicking on a hyperlink. Consequently, those who may not have previously supported a campaign can do so with minimal effort.
- **Versatility:** despite its worldwide presence, the web has an extraordinary ability to be tailored to individuals' needs. While offline campaigns

tend to focus on a selection of tactics, often based on practical considerations, the internet enables individuals to be targeted according to the communication tactic that most suits them personally, be it a text, image, report in PDF or link to Instagram, Facebook, Twitter or YouTube. Again, this increases the likelihood of an individual supporting a campaign.

- **Dissemination:** the capacity of messages to 'go viral' is phenomenal. A single email, Facebook post or YouTube video has the potential to be seen by millions within just minutes of having been posted. Messages posted within specific networks have the benefit not only of reaching millions, but of reaching the specific target audience very efficiently.
- **Information:** unlike their print equivalent, documents can be posted on websites at little or no cost and in considerable numbers. Effectively there is no limit to the amount of information a campaign may include. Planning applications and local authority planning documents can be accessed at the touch of a button.
- **Speed:** one of the greatest advantages of online communication, the speed by which a message can be communicated online is considerable. This results in campaigns gaining support extremely quickly.
- **Cost:** at little or no cost, there is no limit to the number of online campaigns, resulting in campaigns existing where they may not have done previously.
- **New balance of power:** largely as a result of low cost, the internet breaks down the perceived asymmetry between public bodies and the general public. Often individuals or small-scale campaign groups are more agile and less risk averse than larger organisations are and as such are more effective in executing an online campaign.
- **Mobilisation and coordination:** the internet facilitates contact between individuals who share common interests and enables them to coordinate joint actions. It also has the potential to facilitate the formation of new political and social forces which may previously have been hindered by practicalities and resources. Powerful communities of interest can be formed regardless of geographical and social constraints.
- **Different dialogues:** with internet communication offering dialogue in the form of one to one and one to many, the appropriate form of dialogue can be selected and used effectively. Furthermore, petitions could be considered a form of 'many to one' and as such are a particularly powerful voice. Common to each of these forms of communication is two-way dialogue, which enables campaigns to grow quickly, also offering opportunities for more proactive developers to enter into dialogue with potential objectors.

- **Debate and discussion:** it follows, therefore that debate and discussion can occur more easily on the internet than elsewhere. Online communication is an ecosystem founded on interconnected conversations, and in many cases a campaign can benefit from positioning itself on an existing platform, such as that of a popular local website or blog.
- **Individuality:** despite the potential to collate support, many internet campaigns are initiated by an individual, because of the efficiencies afforded to them. A single point of view, if well timed and irrespective of the weight of popular opinion, has the potential to form a powerful campaign. Similarly, the internet enables campaigns to take place on a 'hyperlocal' level, as the next section demonstrates.
- **Low key:** today's activism need not be led by powerful personalities or instigated with great panache; in fact many online campaigns are anonymous. This brings about a lack of accountability which can distort a campaign and present difficulties for the organisations to whom the campaign is aimed.

It goes without saying that online campaigns are potentially more sophisticated, informed, effective, efficient, adaptable, egalitarian and flexible than those that went before them. However, the use of the internet brings about new issues and concerns, one of which is the potential for misinformation. While the internet increases the opportunity for access to information, transparency and accountability, most websites lack the editorial filter that is an important part of professional news generation. It becomes the responsibility of users themselves to assess the veracity of information found online, but where this fails to happen, inaccurate information can be spread too easily.

Furthermore, campaigning has not shifted from offline to online: offline campaigns remain, and they remain successful (often *because* they are supported by online campaigns). This presents additional challenges to developers and therefore a need to understand how online campaigning works and understand the appropriate time to engage with a campaign, and do so effectively.

The internet and local communities

The rise of the 'hyperlocal' website is of critical importance to the planner, developer and local authority. Not only do hyperlocal websites play an extremely constructive role in promoting and debating local issues, they have considerable campaigning potential and as such warrant an understanding by the profession.

The term 'hyperlocal', which originates from the US, describes online local services, which are usually run by local communities and residents' groups.

Box 5.5 Typical content of hyperlocal websites

News

- Community news
- Sports news
- Events (meetings of local clubs or societies, community celebrations, key council meetings)
- News of planning decisions or disputes, local planning and neighbourhood planning developments
- News from local courts, police and schools
- News submitted by local residents
- Articles by local residents

Political

- Election coverage
- Borough and parish council news
- Details of MPs, councillors and candidates

Engagement

- Discussion forums
- Information about local authority consultations

Networking

- Information about local groups including residents' associations
- Information about local businesses

Campaigns

- A range of local campaigns, often concerning local authority services, planning or construction work

Comment

- Blogs

Images

- Photo gallery

Information

- Bus timetables
- Local guides
- Waste collection information

Hyperlocal websites tend to fill the gap left by local newspapers, and thus are both functional, informing people of local news and information, and also emotional, in giving people a sense of local belonging. They provide a new means whereby people can form an attachment not just to their city, town or village, but also to their neighbourhood and street.

Hyperlocals may be run by individual bloggers, small businesses or, in the case of Streetview and About My Area, national organisations. They each have in common the aim of improving the provision of local news, providing information and increasing opportunities for members of their communities to connect.

Williams, Barnett, Harte, and Townend (2014)[11] have produced some very thorough research into the emergence of hyperlocal websites. In a comprehensive survey which explored, among other subjects, the reasons for forming a hyperlocal website, they identify that approximately 70% are instigated on the basis of active community participation; over half see their service as local journalism, and over half as an expression of active citizenship. In terms of online activity, nearly three-quarters had covered local campaigns instigated by others, while over a third had instigated campaigns themselves. This frequently involved holding local authorities to account or forcing democracy in innovative ways. Campaigning tended to focus on failures by service-providers, the need for environmental improvements, cuts to local public services, improvements to local amenities, and local council accountability and planning disputes.

Campaigns around planning tend to focus on large-scale developments, protecting green spaces or protecting local businesses challenged by national chain stores. Campaigning for improvements to local infrastructure was also common, particularly in relation to local roads, train lines and cycle paths.

Aspects of Localism including neighbourhood planning, Community Right to Build and Community Asset Transfer frequently feature on hyperlocal websites, though Williams, Barnett, Harte, and Townend determine that 'The numerous campaigns mentioned against these . . . suggest concern in some communities about how democratic they actually are in practice.'

With so many online services (some of them blogs and Facebook pages rather than websites) coming under the rarely used category of hyperlocal, it is difficult to ascertain how many exist. A UK-based website, Local Web List[12] summarised, in September 2016, that there existed 546 in England, 2 in the Isle of Man, 3 in Northern Ireland, 5 in the Republic of Ireland, 63 in Scotland and 47 in Wales.

Table 5.1 The focus of campaigns initiated and supported by UK community news
producers (coded from 285 qualitative responses)

Focus of campaign	Most recent initiated	Most important initiated	Most recent supported	Most important supported
Planning/Licensing	13	14	30	19
Local public services	11	11	23	18
Improvements to infrastructure	8	7	18	6
Local business issues	6	5	13	9
Local charity	2	1	9	2
Environmental issues	7	3	4	7
Community action	4	2	4	4
Improvements to amenities	10	12	3	5
Council (other)	3	2	2	5
Council accountability	11	4	1	1
Council malpractice	0	2	1	2
National political issues	3	0	1	1
Other	8	6	9	4

Source: Williams, Barnett, Harte, and Townend (2014)

Table 5.2 The focus of investigations carried out by UK community news producers
(coded from 100 qualitative responses)

Focus of investigation:	Most recent investigation	Most important investigation
Planning/Licensing	10	9
Local health/social care services	6	3
Council malpractice	6	7
Council cuts	6	6
Local infrastructure	6	6
Council accountability	5	5
Local amenities	4	2
Local business issues	4	2
Local education services	3	2
Local police	3	1
Environmental issues	3	3
Council waste	2	0
Other	4	2

Source: Williams, Barnett, Harte, and Townend (2014)

The research by Williams, Barnett, Harte, and Townend reveals that there
is a broad distribution of audience sizes amongst UK hyperlocal sites, with
the great majority of sites having relatively small audiences. Just two of the

websites surveyed claimed a monthly average of over 100,000 unique users, while 33 claimed between 10,000 and 100,000 and the remaining 55 below 10,000.

The future of hyperlocal sites is unclear. Some of the most proactive sites have already burned themselves out, overwhelmed by information and opportunities to campaign but struggling to do so with only volunteers to run the service. Those running and using the sites have high expectations for future development. This includes the use of GEO RSS feeds to provide local information via an app, increased use of video and audio, and pressure to carry out investigative reporting and use of Freedom of Information requests. The BBC, the *Guardian* newspaper and the government's Technology Strategy Board have made tentative steps to preserve local sites, but in many cases this has been met with objections on the basis that the raison d'être is that of community ownership.

On the other hand, as efficiencies increase, simple template websites become more readily available, and as the retired generation becomes increasingly IT literate, their potential impact on development is likely to increase further. There is no doubt that producing around 2,500 news stories a week across the UK,[13] hyperlocal websites have a role to play in planning and ongoing dialogue between a developer and a community.

Proactive use of the internet by the development industry

Slowly but surely, the development industry has accepted that the internet has a huge influence on planning, perhaps more so than any piece of legislation ever has. And unlike legislation, the surge in online communication cannot be reversed.

As developers increasingly appreciate the benefits of online consultation, usage generally falls into one of three categories:

- Consultation via social media: to date the most popular tactics for consultation are Facebook and Twitter, but LinkedIn, RSS feeds, YouTube, Flickr and Instagram are also commonly used by both developers and local authority
- Consultation using 'off the shelf' consultation websites such as Citizen Space, E consult and Community Tools
- Consultation using specially developed consultation websites which might be put in place by the consulting organisation (sometimes as microsite attached to an existing website) or set up by a specialist company such as ConsultOnline – perhaps with additional tools, such as Sticky World – embedded in the website.

Box 5.6 Social media usage in the UK 2016

According to figures published by Ofcom in 2016,[14] more than seven in ten internet users have a social media profile – a figure unchanged since 2014. When expressed as a proportion of all adults (rather than all internet users) this figure falls to around six in ten (63%).

While a majority of internet users aged 16–24 (91%), 25–34 (90%), 35–44 (81%) and 45–54 (74%) have a social media profile, this compares to half of those aged 55–64 (51%) and 30% of those over 65. Since 2014, there has been no change in the age group more or less likely to have a social media profile.

Of those adults with a social media profile, no fewer than 95% choose to use Facebook. More than a quarter of social media users have a profile on WhatsApp (28%) or Twitter (26%), with around one in five using YouTube (22%) or Instagram (22%). More than one in ten have a profile on LinkedIn (14%), Snapchat (12%) or Google+ (11%). Since 2014, decreasing numbers use Facebook (95% vs. 97%) but more are likely to have a profile on YouTube (22% vs. 17%), Instagram (22% vs. 16%) and Snapchat (12% vs. 9%). The results show some interesting patterns in terms of demographics:

- Compared to all adults with a profile, those aged 16–24 are more likely to have a profile on five of the ten social media sites/ apps shown above: Instagram (47%), WhatsApp (39%), Twitter (38%), Snapchat (37%) and YouTube (36%). This age group are less likely to have a profile on LinkedIn (8% vs. 14%). Adults aged 25–34 are also more likely to have a profile on WhatsApp (39%) or Instagram (29%).
- Those aged 35–44 and 45–54 are less likely to have a profile on Instagram (14% for 35–44 and 11% for 45–54) or Snapchat (5% and 2% respectively).
- Those over 55 are less likely to have a profile on Twitter (15%) WhatsApp (12%), YouTube (10%), Instagram (4%), Google+ (3%), Pinterest (2%) and Snapchat (1%).
- AB adults are more likely to have a profile on Twitter (343%) and LinkedIn (28%), while those in DE[15] households are more likely to have a profile on Facebook (99%) and less likely to have a profile on Twitter (19%) or LinkedIn (5%).
- Women are more likely than men to have a profile on Facebook (98% vs. 92%) or Pinterest (10% vs. 4%), while men are more likely than women to have a profile on LinkedIn (17% vs. 12%) or Google+ (14% vs. 9%).

Social media is extremely helpful in promoting consultation websites, but less effective in hosting a consultation as Table 5.3 shows.

However, in ideal circumstances, a combination of a consultation website and social media will be deployed. Developers and local authorities

Table 5.3 Facebook and Twitter: a comparison

	Facebook	*Twitter*
Benefits	Speed – a Twitter profile, Facebook Group or Page can be set up in minutes and all information posted will be communicated immediately.	
	Expense – time is the only cost (though may be excessive if the messages do not naturally accumulate interest).	
	Popular with young people, providing the opportunity to encourage them to spread the message within their specific communities.	
	Information posted appears immediately in users' Facebook/ Twitter feeds – information will be received without the user specifically accessing the relevant website.	
	The emerging use of location tags enables a level of local dialogue not previously possible.	
	A single message can quickly gather interest if 'liked', 'shared' or 'retweeted'.	
	The use of Facebook and Twitter widgets on existing websites or blogs is a quick and effective means of drawing potential users to the page.	
	Ability to communicate both in a public group or one to one.	
	Facebook advertising enables a message to be targeted to a specific demographic and geographic area.	Hashtags (#) enable the consultor to join an existing conversation simply by using the hashtag before the key words.
	Ability to form a 'group' of those interested in a scheme and keep them updated through their chosen means of communication.	Live tweeting, such as at a consultation event or council meeting, can draw a wide audience (both those attending the event and those unable to do so) and introduce a new level of dialogue.
	It is possible to ascertain levels of support by encouraging 'likes' for either specific ideas or the project as a whole.	
	Posting in specific interest groups enables the message to spread through communities of interest.	

(Continued)

Table 5.3 (Continued)

		Facebook	Twitter
		Messages can be spread effectively by targeting those likely to share an interest in the scheme by messaging them or posting on their walls.	
Functionality		Post news	
		Post polls	
		Post images	
		Post videos	
		Initiate online forums	
		Arrange events	Connect into existing discussions by using the hashtag (#)
Limitations		Requires user to have, or to create, a Facebook or Twitter profile.	
		The platform remains the property of Facebook/Twitter and as such could change without prior notice.	
		Analysis is limited to Facebook/Twitter's standard analytics which does provide the desired information – for example in relation to the location of those commenting. Although good diagnostic tools are available, they fail to provide the information necessary for a consultation report, and there are no easy means of combining research from Twitter and Facebook, let alone collating this information alongside the more sophisticated level of demographic/geographic data available through a dedicated consultation website.	
		Facebook pages do not rank highly in Google searches, and therefore it is difficult for potential consultees to find a relevant Facebook Page or Group through a generic search.	The content of tweets is restricted by the 140 character limit.
Best used for		Consultations by organisations which already have a strong Facebook/Twitter presence and therefore the opportunity to build on existing support	
		Building interest within existing Facebook communities.	Short calls to action.

usually find that owning and managing a website provides the necessary control and flexibility to host all plans, images and information clearly, to enter into dialogue at an appropriate level, and to both monitor usage and gain an understanding of sentiment. The best consultation websites are also designed in such a way that results are produced in a style and format relevant to the consultation report that will be a necessary component of the planning application.

The communications theorist Grunig[16] defined excellence in communication as that which promotes the use of research, dialogue and consultation to manage conflict, improve understanding and build constructive relationships with a wide range of publics. Online consultation is capable of addressing these principles:

- The internet is by far the most powerful research resource. A substantial proportion of information that is required in researching stakeholder groups and necessary background information is freely and readily available.
- Online consultation allows for real-time dialogue and consultation through a variety of means. Voice recognition, for example, is breaking down barriers and enabling people to communicate online in the way in which suits them best.
- Conflict or crisis/issues management is frequently managed online. There many instances in which this process moves offline, but the internet is probably the single most important tool in managing conflict.
- Creating a constructive relationship is based initially on knowledge, which is best sourced online; similarly, relationships can be formed and developed entirely online.
- A wide range of publics is best identified online – initially. The internet may supply up to 90% of the stakeholder information required for a consultation, but the remaining (and very important) element is often best addressed through personal contact.

Conclusion

As many of today's consultations show, online consultation is extremely successful in practice as well as in theory. Those developers and local authorities which have put in place consultation websites invariably reap the benefits. And as all demographic groups increasingly communicate online and hyperlocal websites and those of special interest groups continue to flourish, the need for developers and local authorities to have a proactive online presence will increase.

Notes

1 Office for National Statistics (2016) *Internet Access – Households and Individuals: 2016* [Online]. Available www.ons.gov.uk/peoplepopulationandcommunity/householdcharacteristics/homeinternetandsocialmediausage/bulletins/internetaccesshouseholdsandindividuals/2016 [Accessed 19 September 2016]

2 Office for National Statistics (2016) *Internet Users in the UK: 2016* [Online]. Available www.ons.gov.uk/businessindustryandtrade/itandinternetindustry/bulletins/internetusers/2016 [Accessed 19 September 2016]

3 Office for National Statistics (2016) *Internet Access – Households and Individuals: 2016* [Online]. Available www.ons.gov.uk/peoplepopulationandcommunity/householdcharacteristics/homeinternetandsocialmediausage/bulletins/internetaccesshouseholdsandindividuals/2016 [Accessed 19 September 2016]

4 Ibid.

5 Ibid.

6 Clark, J., and Aufderheide, P. (2009) *Public Media 2.0: Dynamic Engaged Publics.* Washington, DC: Center for Social Media.

7 The recommendations of Race Online 2012 can be viewed online: www.gov.uk/government/uploads/system/uploads/attachment_data/file/60993/Martha_20Lane_20Fox_s_20letter_20to_20Francis_20Maude_2014th_20Oct_202010.pdf

8 Curtis, S. (2014) *Digital by Default: Why Digital Government Can't Wait* [Online]. Available www.telegraph.co.uk/technology/news/11059238/Digital-by-default-why-digital-government-cant-wait.html [Accessed 19 September 2016]

9 *How Obama's Internet Campaign Changed Politics* (2008) [Online]. Available bits.blogs.nytimes.com/2008/11/07/how-obamas-internet-campaign-changed-politics/?_r=1 [Accessed 19 September 2016]

10 Warman, M. (2010) *General Election 2010: Never Underestimate the Power of the Internet* [Online]. Available www.telegraph.co.uk/news/election-2010/7561876/General-Election-2010-never-underestimate-the-power-of-the-internet.html [Accessed 19 September 2016]

11 Williams, A., Barnett, S., Harte, D., and Townend, J. (2014) [Online]. Available https://hyperlocalsurvey.wordpress.com/

12 http://localweblist.net/http://localweblist.net/

13 *The Future's Bright – the Future's Local* (2014) [Online]. Available www.communityjournalism.co.uk/blog/2014/12/08/the-futures-bright-the-futures-local [Accessed 19 September 2016]

14 Ofcom (2016) *Adults' Media Use and Attitudes Report 2016* [Online]. Available http://stakeholders.ofcom.org.uk/binaries/research/media-literacy/adults-literacy-2016/2016-Adults-media-use-and-attitudes.pdf

15 The social grade descriptors A, B, C1, C2, D, and E are a socio-economic classification produced by the UK Office for National Statistics. AB refers to higher and intermediate managerial, administrative, and professional occupations and DE to semi-skilled and unskilled manual occupations, the unemployed, and lowest grade occupations.

16 Grunig, J. (1982). Excellence in Public Relations and Communications Management. Hillsdale, NJ: Lawrence Erlbaum.

Part II
The planning process

6 The planning process and the role of consultation

Introduction

The legislative framework of the land use planning system in the United Kingdom operates differently in England, Scotland, Wales and Northern Ireland, but the common thread that exists in all four countries is that the planning system should be 'plan-led'.

In practice, this means that at the different tiers of the planning system, from national to the newly implemented neighbourhood level, decisions on whether diverse forms of development activities should receive planning permission or not hinges on compliance with the policies in strategic planning policy, unless there are other 'material considerations' which need to be taken into account.

Where there may be non-compliance with the relevant plan, the weight to be accorded to these material considerations is often a matter of judgement and can be sometimes subjective. As we shall examine further in Chapter 11, a view on where the balance lies between the applicable material considerations is often central to the debate which takes place among elected members of planning committees when determining planning applications.

The plans themselves must also comply with national planning policy and guidance. In England, the 2012 National Planning Policy Framework (NPPF) sets out English planning policy and how it should be applied. Local and neighbourhood plan making as well as development control decisions by local planning authorities need to pay full regard to the NPPF. Inevitably, a tension will always exist between the different tiers of the planning system.

The legislative framework is different in Wales, with the National Development Framework introduced by the Planning (Wales) Act 2015, but similar principles apply with national spatial planning policies cascading down to bodies charged with strategic and local plan making. Northern Ireland produces its Regional Development Strategy and the devolved Scottish Government sets out Scottish Planning Policy.

The common objective is to provide clear and consistent principles and policies upon which planning can function within a plan-led system.

A plan-led system also emphasises the general need for community involvement and public consultation to be woven into each tier from plan making through to decisions on planning applications. As Chapter 8 will show, the recent and continuing emphasis of the current UK government – particularly in England – is to devolve greater power to local communities in taking control and ownership of spatial planning.

This empowerment of the lowest tier of the planning system reflects the commitment by the Conservative Party to 'Let local people have more say on local planning and let them vote on local issues.'[1]

The case for community involvement in planning

Aside from a political impetus that believes that those most affected by development should have the greatest say (although there are limits to this given the hierarchy of the planning system and also planning policy, which has in itself been the subject of litigation),[2] what is the rationale for community involvement and public consultation in the planning system?

Intuitively, it makes perfect sense to involve local communities and other local stakeholders in planning on the basis that local consultations are likely to provide the best environment for good decision-making. Disseminating information about plan making or planning applications and ensuring that local people are aware of the thinking behind policies or proposals, as well as giving them the opportunity to give their views and input, is likely to offer the best prospects of building consensus and can lead to refined or improved planning policies and development proposals.

This 'front-loaded' approach is recommended by the National Planning Policy Framework, which states,

> Early and meaningful engagement and collaboration with neighbourhoods, local organisations and businesses is essential. A wide section of the community should be proactively engaged, so that local plans, as far as possible, reflect a collective vision and a set of agreed priorities for the sustainable development of the area, including those contained in any neighbourhood plans that have been made.[3]

The same applies to planning applications:

> Early engagement has significant potential to improve the efficiency and effectiveness of the planning application system for all parties. Good quality pre-application discussion enables better coordination between public and private resources and improved outcomes for the community.[4]

Although involving more people in and 'democratising' planning may be considered a valuable and worthwhile objective in itself, probably the critical issue is whether it is also achieving 'good' outcomes.

Box 6.1 The benefits of community engagement in the planning process

Enhanced consultation and intensive community engagement in planning is carried out for a number of reasons:

- Avoiding misconceptions about development
- Achieving buy-in
- Building confidence in and awareness of the system, including the relative merits of different options
- Breaking down the perception that planning is decided by people far removed from or not representative of the local area
- Bringing together key players and stakeholders in development and achieving positive relationships, breaking down the 'us and them' mentality that has affected development historically
- Targeting hard-to-reach groups in the community and giving a voice to those that otherwise would not have one
- Helping people understand, identify and ultimately meet needs in their local areas
- Fostering a community-based holistic approach which draws together the economic and social requirements of a community and linking this to funding/investment
- Empowering people to influence and shape the places where they live
- Using local knowledge and expertise to make plans or proposals even more fit for purpose
- Increasing the accountability of elected politicians or officials
- Increasing acceptance of development or growth
- Reducing conflict and potentially speeding up decision-making
- Delivering more development than would otherwise had been provided by a 'top-down' approach.

A question persists, however, as to whether many of these benefits or objectives of community involvement in planning are being achieved in practice.

As we examine further in Chapter 8, although it is easy to point to the popularity of neighbourhood planning simply by referencing the successful backing of every neighbourhood plan put to a referendum to date, it could be equally argued that there is little reliable evidence to suggest that greater local or 'grassroots' involvement in planning has resulted in a greater likelihood to embrace development or indeed delivered more of it than would have resulted otherwise. It is often the case that individuals, community-based groups and politicians embark on defensive or preventative strategies in relation to policy formulation or individual planning applications, and it is a very common and intrinsic feature of the planning system that objectors organise to oppose development.

It could be further argued that the roots of the recent approach of increasing community involvement and consultation in planning were a direct product of a political narrative which was developed in the lead-up to the 2010 general election and could be susceptible to a criticism that it is a 'closed loop'. As David Halpern, of the now privatised Behavioural Insights Team (formerly part of the Cabinet Office) has said,

> Many areas of government have not been tested in the form whatsoever. They are based on hunch, gut feel and narrative. The same is true of many areas outside government. We are effectively flying blind, without much of a clue as to what really works, and what doesn't. It is actually quite scary.[5]

In this respect, community involvement in planning could benefit from the random control testing (RCT) which has been advocated by the Behavioural Insights Team. For example, the results of neighbourhood or micro-level planning could be compared and contrasted with the results achieved by more traditional methods of local plan making in comparable situations within the same or similar local authority.

One of the problems with policy initiatives that have emerged from strong political narrative is that it is often difficult for those who advocated these initiatives in the first place to admit that they may have got things wrong. It would be an admission that the narrative – in this case that local communities are best placed to take decisions around planning and development – is flawed and potentially not meeting the social and economic needs of the locality and ultimately the country as a whole.

However, the current UK government clearly believes that recent reforms to empower the micro level are working and made the claim in May 2016 that 'Plans for housebuilding are more than 10% higher in the first areas with a neighbourhood plan as opposed to only the council's local plan.'[6] It is not entirely clear on what basis this claim was made, but it was used as one of the prime justifications for the further 'localisation' of planning in introducing a Neighbourhood Planning and Infrastructure Bill in the 2016 Queen's Speech.

Community involvement and public consultation in planning in practice

As we have already explored, the system cascades from the macro to the micro level, and there are opportunities within each tier for community involvement and public consultation.

We will explore how this works with local plan making (Chapter 7), neighbourhood planning (Chapter 8) and the determination of individual planning applications (Chapter 10).

Consultation on planning policy

However, at this stage, it is also worth considering the opportunities that exist for individuals and organisations to input into and influence the highest tiers of planning decision-making. At the top, the government operates a dynamic process of introducing new legislation, regulations and guidance.

The government's integrated website GOV.UK regularly features consultations on various aspects of the planning system. The form of the consultation is usually to set out the background of a particular issue or policy and then to invite responses to questions from interested parties.

A typical example was a 'Technical Consultation on Planning'[7] which ran from July to September 2014. This canvassed views on various proposals to improve the planning system and the responses of the government to the consultation were published in documents issued between November 2014 and May 2016 (although this may not have been entirely compliant with the commitment to respond to consultations in a 'timely' fashion – see the UK government's consultation principles below).

In its response to a particular section of the consultation on permitted development rights to increase housing supply (which largely focused on whether the need to apply for planning permission for a change of use to residential should be removed for certain land uses such as offices), the Department of Communities and Local Government (DCLG) noted some 943 responses from a variety of stakeholders, including local authorities, members of the public, developers, businesses, community groups and sector representatives.

The response stated that 'It should be noted that in reaching decisions, the Government was particularly interested in the issues raised, and consequently did not reach a view based solely on the absolute level of support.'

In other words, 'this was not a referendum'! The government response then set out what measures it had undertaken.

Within the lawmaking functions of government, there are also opportunities for individuals and organisations to influence legislation, although in practice this is largely the domain of organisations, politicians, lobby groups and lobbyists who possess the resources and expertise.

The passage of a bill through the UK Parliament offers opportunities for external stakeholders to influence the precise wording and content of proposed legislation and activities to achieve this took place with limited success during the eight months' consideration of the Housing and Planning Bill in 2015 and 2016. There are possibilities for individuals and organisations, for example at the committee stage in the House of Commons, to present evidence to try and amend bills, as this is the point when every line of proposed legislation is examined. In November 2015, various representatives of local authorities, trade associations, professional bodies and housing associations were involved in the process.

Ultimately, the control of a bill lies with the government, and so it proved with the Housing and Planning Bill, which despite some attempted 282 amendments from the House of Lords, went through to enactment with very few changes.

Box 6.2 The UK government's consultation principles 2016[8]

The Cabinet Office sets out the consultation framework for government departments and has established a number of guiding principles that should be followed.

Consultations should be clear and concise

Use plain English and avoid acronyms. Be clear what questions you are asking and limit the number of questions to those that are necessary. Make them easy to understand and easy to answer. Avoid lengthy documents when possible and consider merging those on related topics.

Consultations should have a purpose

Do not consult for the sake of it. Ask departmental lawyers whether you have a legal duty to consult. Take consultation responses into account when taking policy forward. Consult about policies or implementation plans when the development of the policies or plans is at a formative stage. Do not ask questions about issues on which you already have a final view.

Consultations should be informative

Give enough information to ensure that those consulted understand the issues and can give informed responses. Include validated assessments of the costs and benefits of the options being considered when possible; this might be required where proposals have an impact on business or the voluntary sector.

Consultations are only part of a process of engagement

Consider whether informal iterative consultation is appropriate, using new digital tools and open, collaborative approaches. Consultation is not just about formal documents and responses. It is an ongoing process.

Consultations should last for a proportionate amount of time

Judge the length of the consultation on the basis of legal advice and taking into account the nature and impact of the proposal. Consulting for too long will unnecessarily delay policy development. Consulting too quickly will not give enough time for consideration and will reduce the quality of responses.

Consultations should be targeted

Consider the full range of people, business and voluntary bodies affected by the policy, and whether representative groups exist. Consider targeting specific groups if appropriate. Ensure they are aware of the consultation and can access it. Consider how to tailor consultation to the needs and preferences of particular groups, such as older people, younger people or people with disabilities that may not respond to traditional consultation methods.

Consultations should take account of the groups being consulted

Consult stakeholders in a way that suits them. Charities may need more time to respond than businesses, for example. When the consultation spans all or part of a holiday period, consider how this may affect consultation and take appropriate mitigating action.

Consultations should be agreed before publication

Seek collective agreement before publishing a written consultation, particularly when consulting on new policy proposals. Consultations should be published on GOV.UK.

Consultation should facilitate scrutiny

Publish any response on the same page on GOV.UK as the original consultation, and ensure it is clear when the government has responded to the consultation. Explain the responses that have been received from consultees and how these have informed the policy. State how many responses have been received.

Government responses to consultations should be published in a timely fashion

Publish responses within 12 weeks of the consultation or provide an explanation why this is not possible. Where consultation concerns a statutory instrument, publish responses before or at the same time as the instrument is laid, except in exceptional circumstances. Allow appropriate time between closing the consultation and implementing policy or legislation.

Consultation exercises should not generally be launched during local or national election periods

If exceptional circumstances make a consultation absolutely essential (for example, for safeguarding public health), departments should seek advice from the Propriety and Ethics team in the Cabinet Office.

Consulting on planning policy in Scotland, Wales and Northern Ireland

As indicated earlier in this chapter, there are different arrangements in Wales, Scotland and Northern Ireland, although the Welsh Assembly, the Scottish government and the Northern Ireland Executive all take steps to consult on their respective planning policy and guidance.

For example, Northern Ireland's Regional Development Strategy (RDS), now the responsibility of the Executive's Department for Infrastructure, was the subject of a three-month consultation in early 2011.

The process followed was a series of public meetings and the issuing of a consultation document which employed the familiar approach of inviting responses to specific questions about the review of the RDS. The report[9] on the results of the public consultation revealed that public participation was limited, with only 124 attendees at the 13 morning, afternoon and evening events that were organised regionally. Similarly, there were only 129 written responses to the public consultation.

There has been a recent move to transfer planning powers to the 11 local councils in Northern Ireland, and the reform of the planning system enshrined in the Planning Act (Northern Ireland) 2011 has strengthened the framework for community involvement both in terms of the Northern Ireland Executive's responsibilities for determining regionally significant applications and planning policy, as well as the plan-making and planning decision-making powers now the responsibility of the individual local authorities. This has resulted in a Statement of Community Involvement[10] being adopted which underlines the commitment to engage with local communities, which is much more in line with good practice that has been followed for some time in the rest of the UK.

Box 6.3 The commitment to community engagement in planning in Northern Ireland

The core principles behind encouraging engagement in the planning system is set out in the Northern Ireland Executive's Statement of Community Involvement, and the Department for Infrastructure's approach and commitment is stated:

'Engaging communities is an essential part of good spatial planning and for an effective and inclusive planning system overall. As such, the Department shall seek to ensure that the process whereby communities engage with the planning system is clear and transparent so that people understand when and how they can have a say in planning decisions which affect them. For example, the Department aims to involve the community in the planning application process by facilitating early engagement through pre-application discussions on regionally significant developments, informing neighbours of development proposals and alerting the wider community through the Planning Portal and advertisements in the press. The Department will also consider all planning issues raised in representations received.

'It is the Department's intention to undertake a proportionate approach to engagement depending on the nature, scale and complexity of the planning issue raised.

'Furthermore, in the formulation of legislation and strategic planning policy, the Department shall seek to involve the community through a process of providing clear information at the earliest stage in order to encourage effective participation. The Department will undertake proactive and timely consultation and the process will be informative, user friendly, and as inclusive as possible. A key principle will be openness and transparency and the Department will make every effort to engage the community and provide feedback at the end of the process or where appropriate.'

Hybrid bills

The hybrid bill is an alternative route by which planning consent may be granted, and as such offers unique opportunities for public involvement. Hybrid bills are rare, occurring in the case of the Channel Tunnel (1987), Dartford-Thurrock Crossing (1988), Severn Bridges (1992), Channel Tunnel Rail Link (1996) and more recently in the case of the proposed High Speed Rail Link (HS2[11]) and Crossrail.[12]

Hybrid bills provide two key stages in which the public are actively involved. The first is between the bill's introduction and its second reading, where time is allocated to the public for commentary on the environmental statement which forms part of the bill.

In the case of the HS2 Hybrid Bill, the content under scrutiny was overwhelming – there were some 50,000 pages of information related to the bill and its environmental statement (ES). During the consultation process on the ES, 21,833 responses were generated, although over half of these were standard format postcards which had been produced by various lobby groups for use by the public.

Aside from the ES, the hybrid bill process allows for public petitions to be heard following the second reading in both houses of Parliament. In the case of the HS2 Hybrid Bill, this led to almost two years of deliberation by the House of Commons Select Committee before the final report in February 2016.[13]

A total of 2,586 petitions was deposited against the bill and its associated additional provisions (APs). The Select Committee heard nearly 1,600 of these and 800 were withdrawn or did not appear before the Committee. A further 300 petitioners joined with others who were raising similar or identical issues.

The report of the committee also had a number of comments about the hybrid bill process itself, which revealed no small degree of exasperation with the content and prolonged nature of the hearings. While the committee commented on HS2 Ltd's 'mixed record' on public involvement, it also recognised that in the case of major infrastructure works, 'It can be difficult to mollify those whose lives face disruption.'

In many respects, it could be argued that there were few good prospects for consensus building through consultation, when the positions of many of the stakeholders concerned were largely focused on seeking to prevent the proposals coming forward at all. It was always going to be adversarial in nature.

One indication of the confrontational context of the HS2 Hybrid Bill was the amount of time spent by the committee in debating whether the petitioners themselves could demonstrate that one or more of the bill's provisions directly and specially affected them. There were a number of challenges to the standing ('locus standi') of petitioners. This resulted in one instance relating to the bill's Additional Provision 4 (largely concerning a tunnel through The Chilterns) in three days of sittings of the committee and 14 hours of deliberation just to decide whether petitioners should be heard at all.

Ultimately, the committee concluded that,

> Many of the current petitioning procedures and hearing arrangements have been inherited from previous eras no longer fit for purpose. . . . The process requires a huge time commitment from the politicians appointed to the Select Committees which has a severe impact on the duties. Recruitment of those Committees may become very difficult.

The committee added,

> There should be less petitioning, with more focus on serious detriment. Clearer, and authoritative, guidance is needed on what constitutes locus standi – that is, what will result in a right to be heard on a petition. . . . There is simply far too much repetition of the same issues before the Committee. There is a conception, based on our experience, that weighing in with another angle on the same point will help strengthen a case. It does not. If some believe that there is a democratic right for everyone who wants to show up to have their say to repeat issues for as long as it takes, they are wrong. Such a conception does not serve the democratic process.

In other words, the hybrid bill process is seen by the politicians responsible as being too inclusive and as a result leads to repetition and proceedings which are unfocused and unnecessarily prolonged. It underlines, perhaps, that consultation procedures on major infrastructure projects can be used tactically to disrupt and delay, although those most directly affected might argue that this is a good thing and only inevitable if a project has not been sufficiently thought through.

Those seeking to change the HS2 Hybrid Bill were not happy with the process either. The campaign manager for 'Stop HS2' said at the time of the issuing of the Select Committee's final report:

> Two years ago, we had great hopes that the HS2 Committee would see how badly planned HS2 has been and make significant changes. However, after almost two years, next to nothing has been demanded by the Committee, and those changes which have been made are minor. In their summing up almost two years of sittings, the Committee mentioned just five places where they have asked for improvements to the design and construction of HS2, as well as asking for the compensation scheme to work better.
>
> We are incredibly disappointed that the feelings of communities along the route of HS2 have largely been ignored, as HS2 has been railroaded through.[14]

Conclusion

In this chapter, we have explored the plan-led basis of the planning system and the role of and justification for public consultation and community involvement. We have also considered some of the opportunities that exist for stakeholders and the public to be involved in the higher tiers of planning policy formulation and legislation, although those wishing to intervene meaningfully at this level need to have a specialist understanding of and the resources for political advocacy and lobbying, which is beyond the scope of this book. In the following chapters, we will consider how community involvement in planning operates within plan making and the determination of planning applications.

Notes

1 Conservatives.com (2016) *The Conservative Party Manifesto* 2015 [Online]. Available www.conservatives.com/manifesto [Accessed 13 October 2016]
2 See, for example, the legal challenge by DLA Delivery Ltd to the Newick Neighbourhood Plan (East Sussex) where one of the key issues was whether a 'lower tier' neighbourhood plan can precede an NPPF-compliant local plan.
3 DCLG (Department for Communities and Local Government) (2014) *The National Planning Policy Framework*, Paragraph 150. London: DCLG.
4 Ibid., Paragraph 187.
5 Syed, M. (2014) *Black Box Thinking*. London: John Murray Publishers.
6 The Queen's Speech, Briefing Note, May 18 2016: Anon, (2016) [Online] Available www.gov.uk/government/uploads/system/uploads/attachment_data/file/524040/Queen_s_Speech_2016_background_notes_.pdf [Accessed 13 October 2016]
7 Gov.uk (2014) *Technical Consultation on Planning – Consultations – GOV.UK* [Online]. Available www.gov.uk/government/consultations/technical-consultation-on-planning [Accessed 13 October 2016]
8 Gov.uk (2012) *Consultation Principles: guidance – Publications – GOV.UK* [Online]. Available www.gov.uk/government/publications/consultation-principles-guidance [Accessed 13 October 2016]
9 Department for Regional Development, Northern Ireland Executive (2012) *Review of the Regional Development Strategy 2035, Final Report on Public Consultation* [Online]. Available www.planningni.gov.uk/index/policy/rds2035.pdf [Accessed 13 October 2016]
10 Planningni.gov.uk (2016) *Department's Statement of Community Involvement (SCI) | Planning Portal* [Online]. Available www.planningni.gov.uk/index/policy/departments-sci.htm [Accessed 13 October 2016]
11 The High Speed Rail (London – West Midlands) Bill, introduced in 2013
12 Crossrail Bill, introduced in 2005
13 High Speed Rail (London – West Midlands) Bill, Second Special Report of Session 2015–16
14 STOP HS2 – The National Campaign Against High Speed Rail 2 – HS2 – No Business Case, No Environmental Case, No Money to Pay for It [Online]. Available http://stophs2.org/category/news [Accessed 13 October 2016]

7 The formulation of a local plan

Local plans lie at the heart of the planning system in the United Kingdom. They set the scene for acceptable development, because applications for planning permission must be determined in accordance with the development plan unless there are material planning considerations which indicate otherwise.

According to the National Planning Policy Framework (NPPF) which applies in England, 'local plans are the key to delivering sustainable development that reflects the vision and aspirations of local communities.'[1] If proposed development accords with an up-to-date local plan, then the National Planning Policy Framework states that it should be approved without delay.

Further, the NPPF states that local plans should be aspirational but realistic and seek to achieve each of the economic, social and environmental dimensions of sustainable development, avoiding significant adverse impacts and pursuing alternative options which reduce or eliminate such impacts.

The strategic priorities for each individual local plan should be clearly set out and the NPPF describes the key policies which should feature (see Box 7.1). All local plans should identify a vision and framework for development within an identified area, answering such questions as what development is intended, where it should be located, when it should happen and how it can be brought forward.

Box 7.1 The strategic priorities of a local plan

According to the NPPF, local plans should put forward policies which seek to deliver:

- The homes and jobs needed in the area
- The provision of retail, leisure and other commercial development

- The provision of infrastructure for transport, telecommunications, waste management, water supply, wastewater, flood risk and coastal change management, and the provision of minerals and energy (including heat)
- The provision of health, security, community and cultural infrastructure and other local facilities
- Climate change mitigation and adaptation, conservation and enhancement of the natural and historic environment, including landscape.

They should also:

- Plan positively for the development and infrastructure required in the area to meet the objectives, principles and policies of the framework
- Be drawn up over an appropriate timescale, preferably a 15-year time horizon, take account of longer term requirements and be kept up to date
- Be based on cooperation with neighbouring authorities, public, voluntary and private sector organisations
- Indicate broad locations for strategic development on a key diagram and land-use designations on a proposals map
- Allocate sites to promote development and flexible use of land, bringing forward new land where necessary, and provide detail on form, scale, access and quantum of development where appropriate
- Identify areas where it may be necessary to limit freedom to change the uses of buildings, and support such restrictions with a clear explanation
- Identify land where development would be inappropriate, for instance because of its environmental or historic significance
- Contain a clear strategy for enhancing the natural, built and historic environment, and supporting nature improvement areas where they have been identified.

Local plan making in England

Key to the success or soundness of individual local plans is the gathering of an evidence base that can provide a clear and objective assessment of the future needs of an area and the opportunities that present themselves to meet those future needs.

The evidence base that is used to inform the production of a local plan must be robust. In the case of housing, for example, there is a need to demonstrate that the plan meets objectively assessed needs (OAN) for market and affordable housing. This includes the identification of individual sites which can deliver a five-year pipeline supply of housing. Local authorities should also satisfy themselves on an annual basis that a five-year land supply exists.

Local plans should also be a collaborative exercise reflecting early and meaningful engagement with a wide and representative cross-section of the community, including neighbourhoods, representative organisations and commercial interests.

The requirement for local planning authorities to proactively involve local communities and other interested parties in producing local plans is enshrined in statutory legislation and regulations.[2] The approach to be followed by individual local planning authorities has to be set out in a Statement of Community Involvement, a local development document that is required to be deemed 'sound' by an independent planning inspector.

How one local planning authority, Wigan Metropolitan Borough, has approached this in its Statement of Community Involvement (adopted in 2015) is described in Box 7.2.

Box 7.2 Local plan making and consultation: Wigan metropolitan borough's SCI

In its Statement of Community Involvement (revised November 2015), Wigan MBC sets out how the council will consult and involve people in the preparation of local plans. Significantly, the council states that it has kept its SCI deliberately short 'to ensure that as many people as possible read it', recognising perhaps that local plan making is a complex and difficult process which by its very nature creates barriers to participation.

Key commitments by the council include:

- Local plans that set out policies for sustainable development to meet the needs and aspirations of the community and based on up-to-date and relevant evidence about the economic, social and environmental characteristics and prospects of the area and
- Consultation at various stages of plan making, including the 'Proposed Submission' stage, after which the plan submission is made to the Government.

The council also explains in general terms *when* it will consult:

- At *commencement,* asking what the plan should contain and engaging with relevant bodies to identify the issues and the evidence needed
- During *preparation,* as appropriate, when gathering evidence, confirming the issues, developing the options for addressing the issues and then the preferred options/approaches, specifically inviting comments on a draft document during a specified time period and considering them prior to the next stage
- At the *publication* stage, when the proposed submission version of the plan is published for formal representations during at least a six weeks period on the soundness of the plan or whether it complies with legal requirements
- From *submission,* when the plan is submitted to the government, having made any small-scale changes to it in response to comments made at the previous stage, following which an independent inspector will be appointed to undertake a *public examination.* People who have made representations at the previous stage can appear at the examination.

The council identifies whom it will involve in plan making such as:

- *Specific consultation bodies* – organisations specified by law for consultation as appropriate, including those responsible for services, utilities and infrastructure provision, parish councils in and adjacent to the borough, adjoining councils and government departments
- *General consultation bodies* – community and voluntary bodies with an interest in the borough and bodies that represent different racial, ethnic, national or religious groups, disabled persons or persons carrying on business in the borough
- *Elected representatives* – local councillors, members of Parliament and members of the European Parliament
- *The general public* – people who live in, work in or visit the borough or have another interest in the borough
- *Businesses* – those with business interests in the borough
- *Landowners, developers and agents* – those who have a direct interest in future development and have a major role to play in providing the facilities and services the borough needs
- *Duty to cooperate prescribed bodies* – neighbouring councils and other prescribed bodies as set out in law, many of whom are also specific consultation bodies.

The council also describes *how* it will consult:

- Contact with appropriate organisations and individuals directly by email, or by post when it is the only means available
- Publicising consultations by methods such as the council's website and other websites as appropriate, press releases, local newspapers and radio, social media, site notices, meetings, workshops and 'drop-in' sessions
- Making consultation documents available on the council's website, on other websites if appropriate, and at council offices and libraries
- The publication of comments received, or a summary of them, as soon as possible, explaining how they have been taken into account in preparing the plan.

There is quite a degree of flexibility in the way in which local authorities engage with local communities at an early stage of plan production. As long as local authorities notify certain groups and individuals of the preparation of a plan and invite representations as well as follow any consultation commitments enshrined in a Statement of Community Involvement, then they will not be in breach of any requirements.

Key stages in local plan making

The key stages in the development of a local plan and the opportunities within these stages for stakeholders to input their views is set out in Figure 7.1 below.[3]

Individual local authorities tend to describe emerging local plans which are subject to consultation as 'issues and options', 'preferred options' or 'pre-publication'. Local authorities are required to set out the intended production timetable for the local plan in a local development scheme, which should be updated regularly to ensure that consultees and interested parties understand the progress and key milestones in the making of the plan.

It is evident that the process of plan making is complex and can be carried out over an extended period. There is substantial technical analysis undertaken, particularly in developing the evidence base, and this can present significant barriers to public participation, as local residents or civic groups struggle to understand how the whole system works or the key issues which are being considered. For example, given that the most frequently contentious parts of local plans tend to focus on the level of new housing

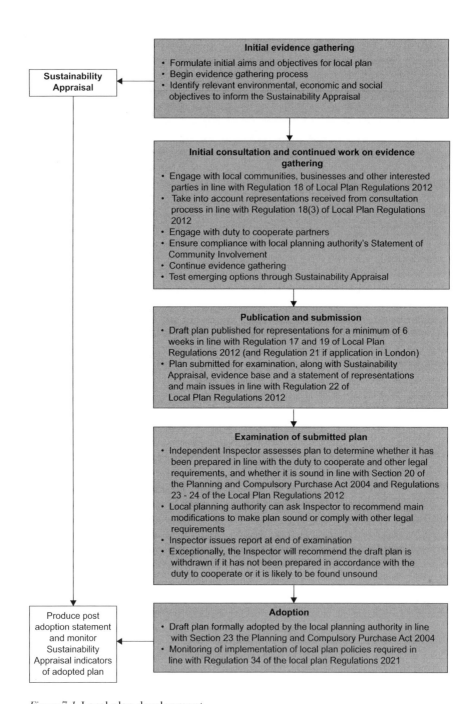

Initial evidence gathering
- Formulate initial aims and objectives for local plan
- Begin evidence gathering process
- Identify relevant environmental, economic and social objectives to inform the Sustainability Appraisal

Sustainability Appraisal

Initial consultation and continued work on evidence gathering
- Engage with local communities, businesses and other interested parties in line with Regulation 18 of Local Plan Regulations 2012
- Take into account representations received from consultation process in line with Regulation 18(3) of Local Plan Regulations 2012
- Engage with duty to cooperate partners
- Ensure compliance with local planning authority's Statement of Community Involvement
- Continue evidence gathering
- Test emerging options through Sustainability Appraisal

Publication and submission
- Draft plan published for representations for a minimum of 6 weeks in line with Regulation 17 and 19 of Local Plan Regulations 2012 (and Regulation 21 if application in London)
- Plan submitted for examination, along with Sustainability Appraisal, evidence base and a statement of representations and main issues in line with Regulation 22 of Local Plan Regulations 2012

Examination of submitted plan
- Independent Inspector assesses plan to determine whether it has been prepared in line with the duty to cooperate and other legal requirements, and whether it is sound in line with Section 20 of the Planning and Compulsory Purchase Act 2004 and Regulations 23 - 24 of the Local Plan Regulations 2012
- Local planning authority can ask Inspector to recommend main modifications to make plan sound or comply with other legal requirements
- Inspector issues report at end of examination
- Exceptionally, the Inspector will recommend the draft plan is withdrawn if it has not been prepared in accordance with the duty to cooperate or it is likely to be found unsound

Produce post adoption statement and monitor Sustainability Appraisal indicators of adopted plan

Adoption
- Draft plan formally adopted by the local planning authority in line with Section 23 the Planning and Compulsory Purchase Act 2004
- Monitoring of implementation of local plan policies required in line with Regulation 34 of the local plan Regulations 2021

Figure 7.1 Local plan development

that is required and its spatial allocation, it is often very difficult for local people to grasp the often complex calculations which arrive at an objective assessment of housing needs for the future.

The problem is exacerbated by the lack of clarity around how housing needs in a local authority area should actually be assessed. The evidence base is supposed to revolve around strategic housing market assessments (SHMAs), yet there is no clear guidance as to how to determine the boundary for SHMAs or indeed how to prepare them. Such uncertainty over probably the most contentious area which has to be addressed by local planning has created significant controversy and difficulties in many areas of the country.

The sheer length of time taken on average to complete local plans is a crucial factor in the under-delivery of housing nationally and impacts negatively on the effectiveness of public participation in emerging planning policy. Statistics provided by the Department for Communities and Local Government (DCLG) and shown in Figure 7.2 demonstrates that local planning is a lengthy process. Consequently in 2016, moves were made by government to potentially intervene and expedite decision-making.

Achieving a balance between providing sufficient detail yet avoiding confusion amongst consultees is a significant challenge, and it is not unusual

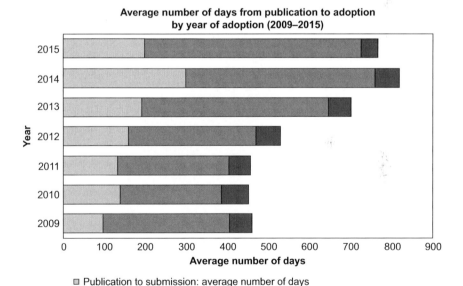

Figure 7.2 Local plan time periods 2009–2015

for local authorities to be criticised for producing local plan documentation that those with little familiarity with the system find difficult to understand. There is a real danger that the system itself has a disempowering
effect on stakeholders because it is divorced from the technical capacity of
consultees and replete with jargon.

This matter was taken up by a pressure group, the Plain English Campaign, in 2015 following complaints that Thanet District Council's 263-page
local plan was 'very hard to understand'. The campaign pointed out that the
consultation included an associated questionnaire, which ran to 99 pages,
and that the district contained a high proportion of residents over 65. The
conclusion was that 'No one is going to, or should be expected to, read 263
badly written pages, or have two screens open in order to answer a questionnaire.' For its part, the council pointed out its legal requirements and
that the local plan was necessarily 'a detailed, technical document which
informs how policy and planning decisions in the future are determined
and importantly has to be approved by the planning inspector'.[4]

Sustainability appraisals

Alongside the gathering of a robust evidence base for the production of
local plans, local authorities are also required to carry out a sustainability appraisal. This should commence at the same time as initial work on
developing the plan, and should systematically consider the plan's wider
economic and social effects as well as its potential impacts. It should also
examine any ways in which potential adverse impacts might be mitigated
and is a critical tool in the early stages of plan making, since it should evaluate the different options or draft proposals in the local plan.

How the sustainability appraisal relates to the key stages of local plan
production is also described in Planning Practice Guidance and is set out
in Figure 7.3.

Where it becomes apparent that either neighbourhood plans or supplementary planning documents could have significant environmental effects,
a separate European Union requirement[5] exists (implemented through
domestic legislation[6]) for a Strategic Environmental Assessment. This aims
to ensure that environmental considerations are fully integrated into the
preparation and adoption of plans and programmes which seek to achieve
sustainable development.

The sustainability appraisal must be the subject of consultation with
defined environmental bodies, such as Historic England, Natural England
and The Environment Agency, as well as other parties which are affected
or which are likely to be affected by any decisions which form part of the
assessment, adoption or making of the plan. The views of consultees on
the local plan itself may also be sought on the sustainability appraisal and
the report relating to it with a separate non-technical summary must be
published alongside the draft local plan.

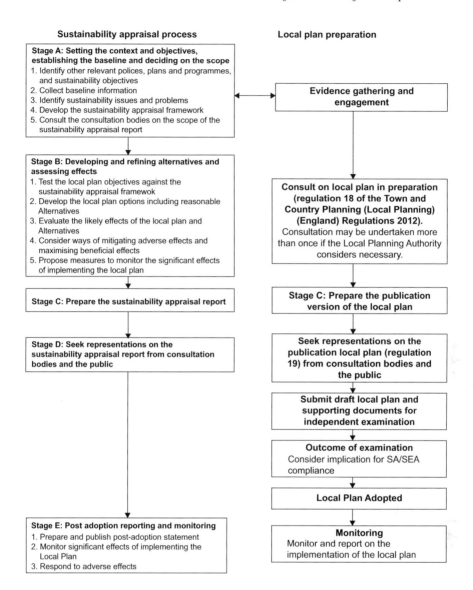

Sustainability appraisal process

Stage A: Setting the context and objectives, establishing the baseline and deciding on the scope
1. Identify other relevant polices, plans and programmes, and sustainability objectives
2. Collect baseline information
3. Identify sustainability issues and problems
4. Develop the sustainability appraisal framework
5. Consult the consultation bodies on the scope of the sustainability appraisal report

Stage B: Developing and refining alternatives and assessing effects
1. Test the local plan objectives against the sustainability appraisal framewok
2. Develop the local plan options including reasonable Alternatives
3. Evaluate the likely effects of the local plan and Alternatives
4. Consider ways of mitigating adverse effects and maximising beneficial effects
5. Propose measures to monitor the significant effects of implementing the local plan

Stage C: Prepare the sustainability appraisal report

Stage D: Seek representations on the sustainability appraisal report from consultation bodies and the public

Stage E: Post adoption reporting and monitoring
1. Prepare and publish post-adoption statement
2. Monitor significant effects of implementing the Local Plan
3. Respond to adverse effects

Local plan preparation

Evidence gathering and engagement

Consult on local plan in preparation (regulation 18 of the Town and Country Planning (Local Planning) (England) Regulations 2012). Consultation may be undertaken more than once if the Local Planning Authority considers necessary.

Stage C: Prepare the publication version of the local plan

Seek representations on the publication local plan (regulation 19) from consultation bodies and the public

Submit draft local plan and supporting documents for independent examination

Outcome of examination Consider implication for SA/SEA compliance

Local Plan Adopted

Monitoring Monitor and report on the implementation of the local plan

Figure 7.3 The sustainability appraisal process

Submission of a draft local plan and the examination process

Having been through the initial evidence gathering stages, consultation and further evidence gathering, the local planning authority then publishes the draft plan which is intended for submission to the Planning Inspectorate for examination. Six weeks are allowed for any further representations on this published version of the plan, and the local authority should provide a statement outlining the representations and main issues to the examination, along with the sustainability appraisal.

The examination process begins with the submission to the Planning Inspectorate and ends when the independent inspector's report is issued to the local authority. The principal task of the inspector is to decide whether the submitted plan meets various legal requirements and is also 'sound'. Soundness is measured by consideration against a number of tests and these are summarised in Box 7.3.

Box 7.3 Tests of 'soundness in local plan making'

When examining a local plan, the inspector will apply a series of key tests to establish whether it is 'sound', and these are contained at paragraph 182 of the National Planning Policy Framework. Plans need to be able to demonstrate that they are:

- **Positively prepared** – the plan should be prepared based on a strategy which seeks to meet objectively assessed development and infrastructure requirements, including unmet requirements from neighbouring authorities where it is reasonable to do so and consistent with achieving sustainable development
- **Justified** – the plan should be the most appropriate strategy, when considered against the reasonable alternatives, based on proportionate evidence
- **Effective** – the plan should be deliverable over its period and based on effective joint working on cross-boundary strategic priorities
- **Consistent with national policy** – the plan should enable the delivery of sustainable development in accordance with the policies in the framework.

There have been a number of examples where an inspector has ruled that local plans are unsound, including diverse local authority areas such as Durham, North Somerset, Maldon, Central Bedfordshire, Hereford, Hart, Chiltern, Canterbury, Aylesbury Vale and Warwick. The usual reason why plans are deemed unsound is as a result of a failure to meet housing needs, meaning that they are not 'positively prepared'.

Failure to cooperate with adjoining local authorities has often been an issue, with territorial disputes taking place between councils over who should address the needs of cross-boundary housing markets.

The loss of the strategic dimension following the abolition of regionally based planning has been a key factor here and not fully addressed by the 'duty to cooperate' (see separate section in this chapter) which emerged from the reform of the planning system.

A March 2015 study by Nathaniel Lichfield & Partners[7] found that of the 43 plans then being examined, 14 had been put on hold, requiring more evidence of objectively assessed housing need, and that 17 plans submitted for examination after the introduction of the National Planning Policy Framework (NPPF) in 2012 had been ongoing for a year or more. Of the 60 plans outside London that had been found sound post-NPPF, the study found that it had taken an average of 15 months to get from submission to being passed as sound.

Given that local plans need to be consistent with the fundamental principles and policies of national policy, it follows that having an up-to-date local plan is highly desirable. However, a report[8] produced by the Local Plans Expert Group, established by government in 2015 to consider how plan making could be made more efficient and effective, indicated that only 31%[9] of the country's local authorities had an up-to-date local plan, and that as a result less than half the country's housing needs were currently being provided for in local plans.

Since the introduction of the Planning and Compulsory Purchase Act 2004, which set out the current basis for plan making in local authorities, some 84% had published a local plan and 68% had adopted a local plan. However, despite guidance setting out clearly that most local plans are likely to require updating in whole or in part at least every five years, at the end of January 2016, only 45% of authorities had a local plan which had been adopted in the last five years, with a further 23% having a local plan compliant with the 2004 Act but over five years old.

This situation has led to the introduction of greater powers in the local plan process for the secretary of state within the Housing and Planning Act 2016, where a failure or omission by a local authority with the preparation, revision or adoption of a local plan could lead to central intervention. The ability of the secretary of state to modify a local plan because it is 'unsatisfactory' has existed in legislation since 2004,[10] but a technical consultation[11] on this issue in February 2016 clarified when such intervention might be prioritised. Triggers for intervention suggested were where there is under-delivery of housing in areas of high housing pressure; the least progress in plan making has been made; plans have not been kept up-to-date; or where intervention will have the greatest impact in accelerating local plan production.

Although undoubtedly motivated by a desire to ensure that local plans are fit for purpose and are not unduly delayed, there is a major question as to how such intervention would be undertaken and how it could be squared with the wider objective of devolving power to the local level. As retired senior planning inspector David Vickery stated,

> How can key decisions affecting an area be done without the guidance, consideration and input of locally elected representatives of the people? Where is the Localism in this? And some NIMBY authorities councillors will be very glad to be able to blame someone else for all the new development.[12]

Local plans in Scotland, Northern Ireland and Wales

As with other aspects of the planning system within the United Kingdom, there are different arrangements for plan making in the constituent countries. However, there are a lot of common elements in the approach followed.

In Scotland, local development plans (LDPs) are published and updated by individual local councils and the National Park Authorities. LDPs provide the policy basis for decisions on planning applications and also identify potential development sites. They aim to cover a 10- to 20-year period, and a 5-year cycle of plan making applies. LDPs are adopted by the planning authority, after submission to Scottish ministers and taking into account the findings of any required examination.

The National Planning Framework (NPF) for Scotland is prepared by the Scottish government. It sets out an overall spatial strategy for the development of Scotland as a whole and defines a number of 'national developments' and other provisions with relevance to Scottish local authorities. Like LDPs, the NPF has to be reviewed every five years.

Additionally, there are strategic development plans (SDPs) which are prepared by the strategic development planning authorities in the four main city regions of Scotland: Glasgow, Edinburgh, Aberdeen and Dundee. They seek to provide a long-term (20 years or more) strategy for the location and delivery of future development. SPDs are approved by Scottish ministers, following examination and any subsequent modification by ministers.

If a planning authority is within a strategic development plan area, then its local development plan must be consistent with the strategic development plan. A local development plan cannot be submitted to Scottish ministers until the relevant strategic development plan has been approved.

In similar fashion to the practice in England where a local development scheme is required, a development plan scheme has to be produced which sets out the timetable for the preparation and review of local developments plans. Within this, a participation statement, required by statutory legislation[13] sets out the opportunities for local people and other stakeholders to input into plan making. Box 7.4 below shows how East Lothian Council carried out consultation during the preparation of its local development plan.

Box 7.4 East Lothian's participation statement

As required under statutory legislation, East Lothian Council published the latest iteration of local development scheme (No. 8) in March 2016. This contained a participation statement which set out each stage of its plan preparation and the related consultation that it undertook.

The council began the process in 2011, and it expects to adopt its local development plan in 2017. For the time being, the current development plan for the local authority is the strategic development plan for Edinburgh and South East Scotland (approved by Scottish ministers in 2013) used in conjunction with the East Lothian local plan (adopted in 2008).

The stages which it followed were:

- Awareness raising (2011), where the intention to prepare the plan was publicised through the media and the council's website. Engagement with community councils, civic groups, business organisations and the development industry also took place and resulted in the creation of a consultation database
- Issues and options (2011–2013), which sought to identify land use options and related planning issues. Here, the council commissioned Planning Aid for Scotland to carry out a series of six community workshop events in the main towns within the local authority. Additional workshops were held with businesses, developers and community councils and a dedicated session held to look at a potential new settlement
- Main Issues Report (MIR, 2014), which puts forward the council's general proposals for development within its administrative boundaries along with specific proposals where development should and should not take place. The MIR has to contain one or more alternative options and also describe how the favoured or alternative options are different from the extant local development plan (if this exists). Here, East Lothian used the media, its own newspaper (delivered to every household), its website, afternoon drop-in sessions and evening workshops (again in its principal towns)
- Additional consultation with community councils (2016), where specific development sites were presented and community council meetings gave their feedback. Four community councils were engaged in this way and members of the public were also involved.

The next step will be to publish the proposed local development plan, which should take into account the representations received on the

MIR. This will be subject to a further six-week period of public consultation. The council's intention is to use the media, its own website and the distribution of the proposed LDP to all the consultees it has identified. The council is able to make changes to the plan, although if significant, these will have to be notified.

Any unresolved objections will be considered at an examination and recommended changes made as a result of the examination will be generally binding on the council. Finally, the LDP, with any modifications coming out of the examination, will be submitted to Scottish ministers and at the expiry of 28 days, the council will be able to adopt the new LDP, unless it receives a direction not to do so by ministers.

Northern Ireland also provides for local councils to produce local development plans. As is the case in the rest of the United Kingdom, the LDP is the primary source for the consideration and determination of planning applications. Until relatively recently, local development planning was centralised within the Northern Ireland government. LDPs are defined in regulations[14] which state that they should have a 15-year time horizon.

LDPs are subdivided between the planned strategy and local policies. The Northern Ireland government has an overarching role in that it will seek to satisfy itself that a LDP is in conformity with central government plans, policies and guidance.

Public involvement is an important part the process, and a Statement of Community Involvement is required, setting out how and when the wider community and relevant stakeholders can be involved.

The stages in plan making are similar to those in England, in that they commence with plan preparation and data collection. A preferred options paper (POP) is prepared, then a draft plan strategy and ultimately an independent examination takes place.

In the 22 Welsh Unitary Authorities and three National Park Authorities, there is also a statutory requirement to produce local development plans. As with the practice in Scotland, the individual LDPs have to take account of national planning policy in Wales.

LDPs cover a 10- to 15-year period and the local planning authority prepares a deposit plan which is a draft of the LDP and reflects the preferred approach of the council in terms of future land use. This deposit plan is subject to an independent examination by the Planning Inspectorate and as in England, the purpose of the examination is to establish whether the plan is 'sound' and that all representations made by interested parties have been fully considered.

Early on in the plan-making process, Welsh local authorities draft a 'delivery agreement', which sets out a timetable for producing the local development plan and also a 'community involvement scheme', which explains how developers, the public and interested groups can contribute to plan preparation. This also explains how responses will be treated and what feedback will be given.

Politics and plan making

There is little doubt that the political dimension in local plan making is very significant. Arriving at an agreed position where new development, particularly new housing, should be located can be highly controversial and as a result politically difficult. Sensitivities can be particularly acute when discussions centre on the allocation of new development to politically marginal parts of the local authority where different political parties are in fierce competition for votes.

A prime example here is in Rochford Council in Essex. Here, opposition to two major new housing developments in the town of Rayleigh led to significant community support being given to a new independent party, Rochford District Residents. Their simple pitch to the electorate was that the ruling party, in this case the Conservatives, was not being responsive to considerable localised concern about and opposition to development pressures. In the local elections of May 2016, the incumbent conservative administration managed to retain control, but saw its majority significantly decreased after losing eight seats, with the Rochford District Residents gaining four of the seats it lost.

The controversy around new development within Rochford also extended into the local plan process and a debate around the sufficiency of infrastructure and the potential loss of green belt land have loomed large. The early revision of the council's core strategy commenced in 2016, with the council working on its evidence base, in particular a strategic housing market assessment. This was actually made public within days of the results of the May 2016 local elections and left the distinct impression that this had been held back because of political sensitivities. This is not unusual: political events and elections often weigh heavily in the time-tabling of local plans and the timing of key decisions or publication of documents.

Rochford Council sought to involve the public through interactive community involvement workshops, including village 'walkabouts', to identify specific needs. However, it meant that as a result of the recent deep-seated divisions about development, the ability to create some kind of consensus was problematic. There appeared to be a fundamental anti-development stance, which was going to be difficult to overcome despite the best efforts of consultation and involvement. In this respect, debates around local plan making are always at risk of becoming polarised, with little room for manoeuvre or compromise. This creates real difficulties for those seeking to engage with and involve the public.

A similar community-based anti-development stance has been evident in the formulation of the Wantage neighbourhood plan, within the administrative area of the Vale of White Horse District Council in Oxfordshire. Here, no land at all was allocated for housing, leading to an examiner concluding that the plan did not promote sustainable development.

The Rochford and Wantage examples are by no means unique and explain why the Local Plans Expert Group in its report[15] has indicated that one of the principal problems facing local plan making is a lack of political will and commitment. It is hardly surprising that politicians lack the motivation to address sensitive issues raised in local planning consultations when they feel that if they attempt to do so they may be punished in local elections.

Often, plan making is simply categorised as 'too difficult' by politicians and there is natural aversion to taking unpopular decisions, particularly when modern communications means that social media, blogs and websites give such decisions even greater exposure as well as heightening the profile of the decision makers. Many campaign groups recognise only too well the pressure that can be exerted on individual local politicians, and it could be argued that this is all part of healthy democracy, reinforcing the accountability of politicians for their decisions. The move towards referendum-based neighbourhood planning has also reinforced the concept of allowing voters to decide on spatial planning rather than elected representatives and gives weight to the notion of 'people power' in plan making.

Box 7.5 Case study: the political realities of local plan making

Tendring District Council, North Essex

It could be argued that the complexity and length of local plan making provides easy opportunities for campaign groups and political parties to launch populist interventions against development. The recent increase in representation of the UK Independence Party (UKIP), whose modus operandi is to focus on single-issue lowest common denominator campaigns, has had implications for plan making.

In Tendring District Council in North Essex, the UKIP platform was blanket opposition to new housing development. The election of a large number of UKIP councillors at the May 2014 local elections meant that much of the evidence-based approach and work that had been followed to that point was placed under increased scrutiny and criticism, with progress on the local plan being stymied.

In 2016, UKIP promised nationally the protection of green spaces and opposition to excessive housing development, wind farms and HS2.[16] Further, the party promised to allocate local homes preferentially to 'local people and veterans first' along with major planning and planning decisions being subject to a binding local referendum if 5% of local people petition for it.

One of the key problems identified by the Conservative leader of Tendring Council, Neil Stock, was that although the council was 'doing the right thing' in pursuing a local plan that would deliver the appropriate number of new houses according to objectively assessed need, this was of little interest to residents who were hostile to new development in principle.

As Councillor Stock, who was also at the time the cabinet member for Planning and also the chair of the Local Plan Working Group, noted,

> We all know the local plan process is long and laborious – perhaps unnecessarily so – keeping the local plan committee enthused and focused is one thing, but keeping residents engaged and willing to participate is another. . . . Although the principles of "Localism" are laudable, the fact is that very few local residents are that interested – in most cases, only those who are opposed to a site will be motivated enough to participate.[17]

On his political opponents, Councillor Stock added,

> UKIP is making decisions on the basis of constituents' views, instead of having a wider strategic view that considers the needs of the community as a whole. . . . But this has led to a disconnect between strategy and delivery. Not only has the local plan working group struggled to move forward, but the planning committee is refusing many of the applications coming before it, even for those sites selected in the emerging local plan, and often against officers' recommendations.

The reality in many parts of England is that new development is not at all popular, and it might be argued that UKIP is only reflecting community opinion. One of approaches employed over recent years has been a move to try and encourage acceptance of new development through the use of financial incentives such as the new homes bonus, where a local authority receives payments over an extended period for each new residential property it permits. However, this scheme does not seem to have had much impact in changing the attitudes to development at a micro level, perhaps because the 'reward' is too detached from individual residents. A recognition that this might be the case has led to new Prime Minister Theresa May trailing the idea in 2016 that payments might be made to individual residents if they were to accept shale gas extraction ('fracking') developments. The potential monetisation of the planning system would bring a whole new dimension to community involvement, which would shift it from a persuasion/consensus-building approach to one more based on negotiation/compensation.

However, aside from financial incentives related to plan making or planning applications ('the carrot'), as has been set out elsewhere in this chapter, the likely alternative is either central government intervention ('the stick') to ensure that growth is delivered or planning by appeal (effectively disengagement from the process by local authorities), which could lead to significant costs being awarded where a planning inspector deems a refusal to have been unreasonable.

Cllr Stock's ultimate take on his own council's experience was that 'After all, having an unpopular local plan is better than having no plan. And consenting to larger applications on the proviso that they meet tight conditions has to be better than losing at appeal and having no control at all.'

There can be no doubt that politics plays a very significant role in plan making, and there is little sign that this is diminishing. Those involved in promoting new development need a thorough understanding of the political environment in order to present proposals in the most convincing and effective way. They also need to be prepared to rebut those that have a pure anti-development stance by highlighting to political audiences the negative economic and social consequences, as well as the loss of local control, that might follow through the intervention of central government or the appeal system.

It is important to recognise that politically driven local authorities will always be likely to choose the 'least worst' local plan allocation seen from their own assessment of their political interests. This might mean that allocations which on the face of it may have better sustainability credentials may not get political support if they are in locations which are marginal in electoral terms or represented by senior or influential local politicians. This might mean that before acquiring land or options on land, steps are taken to gain an understanding of the political and community environment. This 'political due diligence' for the development industry may well be as important as more traditional forms of risk assessment.

Duty to cooperate in local plan making

Following the abolition of regionally based planning after the arrival of the Coalition government in 2010, the solution brought forward to encourage joined-up planning policy was to introduce a new 'duty to cooperate' between neighbouring local authorities. This created a further community and political dimension to plan making.

The idea was that planning issues that transcended local authority borders, such as sequentially located housing growth on the edge of major urban areas, should be addressed. The National Planning Policy Framework is explicit in its expectation that individual local authorities must meet their own housing needs and also the needs of other local authorities if they are within the same housing market. Equally, cooperation is expected in planning for key required infrastructure, such as transport, telecommunications, energy, water, health, social care and education.

Such need for cooperation, although based on coherent thinking in terms of achieving good strategic planning, has been much more difficult to deliver in practice and has led in many instances to local plans being found 'unsound' by inspectors at examinations. It could be argued that the ongoing systemic resistance to strategic planning is a direct consequence of the national policy move in England away from the pre-2010 approach enshrined in regional spatial strategies (see Chapter 8). Cross-boundary cooperation does not sit easily with Localism.

The conclusion of the inspector examining the Central Bedfordshire local plan in 2015 was that the failure to cooperate to the required level with neighbouring Luton Borough Council on cross-boundary strategic housing delivery had rendered it unsound. The relationship between the Labour-controlled Luton and Conservative-controlled Central Bedfordshire had not been friendly for some time, as the former had launched a judicial challenge on the latter's approval of a 5,150-unit urban extension on green belt land. The two authorities were holding meetings and also sharing information, but observers of the process concluded that practical and demonstrable involvement was not achieved.

Whether concrete and substantial cooperation has taken place often becomes a matter of dispute, and local authorities regularly intervene in neighbouring councils' local plan examinations. For example, in August 2016, the planning inspector involved in the examination of the St Albans Strategic local plan stated in a letter outlining his preliminary concerns that,

> Local Planning Authorities are expected to demonstrate evidence of having made every effort to co-operate with regard to issues with cross-boundary impacts. If a Local Planning Authority (LPA) cannot demonstrate that it has complied with the Duty then the Local Plan will not be able to proceed further in examination. The most likely outcome of

a failure to demonstrate compliance will be that the LPA will withdraw the local plan. Based on the submissions and evidence that I have read (for example from nearby LPAs), I am concerned that the Duty has not been met.[18]

Territorial matters often loom large in the thinking of politicians who struggle to justify the concept of meeting the needs of other local authorities to their own electorates. The response that is heard from the public by politicians – particularly in more rural areas on the boundaries of major towns or cities – is 'Why should we have to solve someone else's problem?' Although there is the ultimate sanction of a local plan not being allowed to proceed to adoption, one of the key problems has been that the duty to cooperate does not mean there is a duty to agree.

The report of the Local Plans Expert Group (LPEG) in March 2016 identified the inherent weakness of the duty to cooperate, which one local authority described to the group as a 'duty to chat'. The LPEG also reported that it had identified 'very few examples in which neighbouring authorities have accepted unmet needs from adjoining authorities.'[19]

The conclusion drawn by the LPEG is that lack of cooperation is one of the principal reasons in the failure of the plan-making system in addressing housing needs and worsening affordability due to supply-side deficiencies. The group has suggested ways in which the duty can be better defined and strengthened and also indicated that where local authorities are not working together, there could be central intervention to direct the preparation of a joint local plan. However, this raises once again the possibility of a fundamental conflict with the core principles of Localism, something which lies very directly at the heart of trying to create a successful plan-making system.

The future of community involvement in local plan making

A critical review of how local plan making has operated might conclude that the process, particularly in England, is complex, is overly long and in many respects produces barriers to effective community involvement. It also reveals that there are real tensions between the ambitions for Localism and the need to provide sustainable growth, especially in relation to housing needs, where there is a chronic undersupply with associated direct impacts on affordability.

One of the key problems faced by local authorities has been the tendency for almost constant revision of national policy and guidance, meaning that the basis on which local plan making can proceed is subject to regular change. Public authorities coping with that in terms of available human and financial resources, in an economic climate where public spending is being cut, makes the task even harder.

The complexity and extent of the evidence base required for plan making means that in many instances there are no standard approaches being followed by local planning authorities. The lack of clarity on how to arrive at an objective assessment of housing need causes significant problems, given that the development of new housing tends to be one of the most contentious issues within local communities. It is also difficult to build community consensus through involvement when the evidence base is so technical. In this respect, there is clearly a need to think about how the process can be simplified and shortened, if community involvement is to be successful in the future.

One of the key concerns regularly encountered during community involvement around plan making or indeed during the consideration of planning applications is whether an area can cope with growth in terms of social and physical infrastructure. Recent experience in London, for example, has consistently demonstrated that there are major pressures on primary school provision in several London boroughs. It is very difficult to build support or consensus within local communities for development growth when there is a strong perception of lack of social provision.

One of the tactics that might be adopted in this respect is for local planning authorities to undertake detailed assessments of existing capacity within their administrative boundaries. A holistic approach and explanation to local communities affected by future development that either there is or will be capacity in environmental and infrastructure terms may well be positive for community involvement.

There is a strong case for streamlining the evidence base required for plan making in order to reduce the lengthy periods involved, which means that focused consultation is hard to maintain. Once the whole process extends beyond 18 months to two years, it becomes difficult to maintain involvement and fatigue sets in. Disenchantment and disempowerment become real factors, leading to withdrawal from the process and the easy route to fundamental opposition – 'a plague on all your houses'.

Undoubtedly, for some politicians, the drawn-out process is seen as no bad thing, as it allows difficult political decisions to be delayed. The system is 'politically convenient'. The very onerous requirements of sustainability appraisals are a clear factor in prolonging the time required. Less detailed yet still robust sustainability statements could feature instead, and these could be far more understandable for lay people and civic groups, whom local authorities are trying to bring into the process.

Whilst there are inherent difficulties and constraints with micro plan making (see next chapter on neighbourhood plans), there is a clear case to make local plans far more 'top level' in content, and there have been suggestions by some that they should be restricted in size. Local plans could also take a staged strategic approach, allowing the space for neighbourhood planning to take on more of the detail.

One of the process problems of plan making in England is that once a planning authority has arrived at its draft local plan (Regulation 19[20] stage), it is precluded from making any amendments. Any representations received are simply passed on to the examining inspector, who is constrained to only take them into account if they bring into question the inherent soundness of the plan itself. In order to address the consultation deficit that exists, local authorities usually introduce further stages to the plan process.

On the face of it, it would do much for good practice in community involvement if community representations on a draft local plan could be taken on board by the local authority itself. In this way, the input of the local community would be seen to have more value, with consultees seeing their local authority dealing with their feedback rather than by an inspector whose remit is somewhat narrow and constrained. It would also remove the need for the introduction of further non-statutory consultation stages.

Certainly, new technology and the use of websites and social media could be deployed more successfully to arrive at more successful community involvement. One of the key problems for community involvement is fear of the unknown, and the voluminous and extensive local plan documents that are routinely found on local planning authority websites are a deterrence to anybody who wants to gain an understanding of how their local area might look in the future. It is incredibly difficult for local people to conceive how their local community might look in the post-plan scenario, and this is where 'propositional planning' could come in, providing images or animations of the consequences of proposed plans.

Having recognised the barriers to community involvement and the weaknesses of the system (which is clearly not delivering the required development that the country needs), this chapter has also indicated the very real political dimensions of plan making and the significant impact of electoral sensitivities.

Whilst local communities could be involved more successfully and even greater use of incentives deployed to build support for growth-delivering local plans, the reality is that long-term dysfunction of the system must always raise the prospect of a need for more pressure being leveraged on local planning authorities to deliver.

This implies measures that might mean the threat of central government intervention in individual plans or the direction to a number of neighbouring local authorities to work on joint local plans to overcome deficiencies in the duty to cooperate. Other initiatives might be the introduction of a statutory duty to have an up-to-date plan and a deadline for delivery, with the implication that failure will mean that the local authority could no longer rely on its extant plan.

The fundamental problem with a more centralised punitive ('stick' as opposed to 'carrot') approach is the conflict with the wider drive towards Localism, an issue which we consider further in the next chapter.

Notes

1 DCLG (Department for Communities and Local Government) (2014) *The National Planning Policy Framework*, Paragraph 150. London: DCLG.

2 The Stationery Office Limited (2012) *Town and Country (Local Planning) (England)* Regulation 18
3 Planning Practice Guidance on Local Plans, NPPF 2012
4 Isle of Thanet Gazette, February 2015.
5 DLGC (Department for Communities and Local Government) (2005) A Practical Guide to the Strategic Environmental Assessment Directive. London: DLGC.
6 The Stationery Office Limited (2004) The Environmental Assessment of Plans and Programmes Regulations. London: HMSO.
7 Nathaniel Lichfield & Partners (2015) Signal Failure? A Review of Local Plans and Housing Requirements. London: Nathaniel Lichfield & Partners.
8 Local Plans Expert Group (2016) Local Plans, Report to the Communities Secretary and the Minister of Housing and Planning. London: Local Plans Expert Group.
9 Based on statistics from DCLG and Nathaniel Lichfield and Partners quoted in Ibid.
10 See in particular Section 21 of the Planning and Compulsory Purchase Act (2004)
11 DCLG (Department for Communities and Local Government) (2016) Technical Consultation on Implementation of Planning Changes. London: DCLG.
12 Town & Country Planning, December 2015
13 The Stationery Office Limited (2006) *The Planning etc. (Scotland Act 2006).* London: HMSO.
14 The Stationery Office Limited (2015) The Planning (Local Development Plan) Regulations (Northern Ireland). London: HMSO.
15 Ibid.
16 UKIP (2016) Local Manifesto. Newton Abbot, Devon: Stephen Crowther.
17 Stock, N. (2016) Political Interference in the Planning Process Will Come at a Cost [Online]. Available www.planningresource.co.uk/article/1391040/political-interference-planning-process-will-cost-neil-stock [Accessed 12 October 2016]
18 Planning Inspector to St Albans District Council, 22 August 2016 [letter].
19 Ibid.
20 The Stationery Office Limited (2012) The Town and Country Planning (Local Planning) (England) Regulations. London: HMSO.

8   Neighbourhood planning

The origins of Localism

During the general election campaign of 2010, the then Conservative opposition presented a strong narrative around decentralisation. Although not specifically using the term 'Localism', the manifesto attempted to lay out a vision for handing down power to individuals and local government. The political objective was clear – to paint a picture of Labour's approach in government since 1997 as inherently bureaucratic and distant from the needs of people and the communities in which they lived.

Making politics 'more local' became an integral part of David Cameron's 'Big Society' pitch to the electorate. It was summarised as follows in the Conservative manifesto:

> We need a totally different approach to governing, one that involves people in making the decisions that affect them. This is what we call collaborative democracy – people taking the kind of powers that until now have been exercised only by governments. So we want to pass power down to people – to individuals where we can, but it is not always possible to give power to individuals, and in those cases we need to push power down to the most appropriate local level: neighbourhood, community and local government.[1]

This philosophy played itself out in relation to the Conservatives' perspectives on the English planning system, particularly the Labour government's regional spatial strategies (RSS), which aimed to develop plans at a regional level that would feed down and inform plan making at local planning authority (LPA) level.

This system was particularly unpopular in the English shires, where the largely Conservative-run councils railed against what they perceived to be imposed housing targets from regional assemblies that contained a mixture of appointees from councils and other regionally based bodies. Not being directly elected and also based on arguably arbitrary boundaries, RSSs were considered by many Conservative politicians as undemocratic and nothing much more than a device to 'force' development on local communities.

Although recognising the importance of the planning system in delivering economic prosperity and a sustainable environment, the Conservative manifesto gave its commitment to a new 'open source' planning system and stated that 'This will mean that people in each neighbourhood will be able to specify what kind of development they want to see in their area. These Neighbourhood Plans will be consolidated into a local plan'.[2]

Having constructed a new Conservative-led coalition government following the May 2010 general election, moves were immediately made to revoke the RSSs, with a letter dispatched in July 2010 to all local authorities informing them that there would be a transitional period leading up to the outright abolition of the previous basis of the planning system through the introduction of a new Localism bill.

The Localism Act, which received royal assent in 2011, introduced the new legislative framework for neighbourhood planning. This has been subject to further definition and change through the adoption of various regulations[3] and most recently by the Housing and Planning Act 2016.

The legislation applies to England. There are no equivalent rights afforded to local communities in Wales, Scotland and Northern Ireland, although there are community planning initiatives which apply there.

The May 2016 Queen's Speech also revealed further appetite by the Conservative government for the extension of neighbourhood planning with the announcement of the introduction of a new neighbourhood planning bill. According to a note[4] published to accompany the Queen's Speech, the new bill would 'further strengthen Neighbourhood Planning and give even more power to local people'. The new legislation would place on local government a duty to support groups in achieving greater transparency, and also in improving the process for reviewing and updating plans.

The mechanics of neighbourhood planning

Since April 2012, local communities have been given the power to produce neighbourhood development plans (NDPs) for their local area. The power is vested in parish councils or a specially constituted community-based organisation, usually a neighbourhood forum.

Neighbourhood planning allows for the creation and establishment of general planning policies for the development and use of land in a designated neighbourhood area. The power that is given to the local community is one of being able to shape and influence where new development or growth might go and how it might look.

Linked to the NDP is the neighbourhood development order (NDO), which allows the community to grant planning permission for development that complies with the order. This removes the need for a planning application to be submitted to the local authority. This can apply to major development schemes, retail uses and domestic extensions of a certain scale.

It is important to note from the outset that such neighbourhood planning must pay due regard to the strategic planning context in which the neighbourhood area sits. This means that the local development plan, National Planning Policy and also European Union obligations, including human rights requirements, have to be taken into account; neighbourhood planning cannot operate in a vacuum.

The need for micro-level planning to be compliant with high-level or strategic policy is a matter which has been the subject of recent litigation, and this is examined further below in a case study (see Box 8.1).

As with all plan making, the relevant legislation and regulations set out a number of stages which must be followed in taking forward neighbourhood planning.

Neighbourhood area

The first formal stage in the process is to identify and agree the neighbourhood area itself with the local planning authority. The requirements to take this forward include a map identifying the geographical boundaries, a justification for the designation of the area and a statement from the applicant body setting out its credentials as a qualifying body to transact the proposed neighbourhood planning.

There are no specific guidelines as to how the geographical boundaries should be fixed, although there is a presumption that parish or town councils will use their established boundaries. However, it is entirely possible to focus on town or local centres or even land immediately adjacent to a boundary, as this might be the subject of potential future development or growth. In this respect, a neighbourhood area can cover two or more local authority areas.

If the area is wholly or predominantly businesses in nature, then the legislation states that the LPA must consider whether it should be designated as a business area for neighbourhood planning.[5] This has implications for any eventual referendum (see below), as the plan needs to pass two referendums: one for residents and one for non-domestic ratepayers. This extra referendum provides businesses with an opportunity to vote on the plan.

Where there is no town or parish council, the alternative is to form a special purpose body known as a neighbourhood forum. Requirements for formation of a neighbourhood forum are that it has to include at least 21 people, with a minimum requirement that at least one of the members lives in the area, one works in the area and one is an elected councillor. The expectation is that the membership of the intended forum should reflect the inclusivity, diversity and character of the area. A written constitution is also required.

There are time limits which apply to this first stage of the process, and there was some concern from government that local authorities were being dilatory in ensuring that the neighbourhood planning system was being transacted speedily. Regulations were introduced in 2015 to ensure that decisions were taken on the designation of a neighbourhood area within specified time limits – 20 weeks from when the application is first publicised in the case of an area which extended beyond two or more local authority areas, 13 weeks in a single local authority area or 8 weeks when the area put forward is by a parish council and it relates to the whole of the parish.

Commencement of the plan process and duty to support

Once this initial stage has been concluded, the work of producing a neighbourhood plan can begin in earnest and the relevant local planning authority is legally obliged to provide help and assistance. This 'duty to support' can take a number of forms, including providing policy input, technical advice or expert consultancy advice, making meeting/consultation space available and ultimately organising the independent examination and local referendum.

The involvement of the local planning authority throughout is considered to be very important in order to ensure that neighbourhood and local plans are complementary and that policy conflicts are minimised. This is particularly relevant to the delivery of housing, where the macro needs of a local planning authority need to be met through micro-level delivery. The dynamic nature of housing needs – nothing ever stands still – has also been stressed recently by government in National Planning Practice Guidance, underlining the requirement for neighbourhood planning to set out indicative delivery timetables and reserve sites that could be used in the future to meet local needs. The propensity of neighbourhood planning to be 'defensive' or 'protectionist' in practice is examined further below.

Box 8.1 Case study: Cockermouth neighbourhood development order

The pioneer in neighbourhood development orders was in the Cumbrian town of Cockermouth when, after a three-year process, a local referendum in 2014 approved the order which permitted certain developments in particular areas without the need to submit a planning application.

In particular the Cockermouth NDO permits:

- Commercial properties in the town's historic market place area to convert into cafés, bars and restaurants
- The conversion of the upper floors of commercial properties on two town centre streets into a maximum of four flats per property

- Traditional timber shopfronts to be installed without an application in the same two streets
- Timber sliding windows and doors to be installed in five other residential streets.

One of the primary motivations for the creation of the NDO was the devastating flood which affected Cockermouth town centre in November 2009. The town council was concerned that local businesses affected by flooding were required to submit planning applications for replacement shopfronts which added cost and delay, hindering the recovery process.

One of the more interesting aspects of the consultation process was the intervention of the external consultee English Heritage, which objected to the relaxation of a requirement for planning permission on certain alterations to houses in the town centre, in particular the installation of replacement windows and doors. Despite attempts to find consensus with English Heritage on a design guide, objections were maintained and this led to a scaling back of the ambition of the NDO.

According to Steve Robinson, the senior planner at Allerdale Borough Council who assisted the town council with the process, it was quite difficult for members of the town council with limited planning expertise to navigate the technical aspects of the NDO. One of the attractions of the town council to the NDO was that the order gave control to the local area through its permissiveness – the borough council would have no role.

The value of an NDO in promoting regeneration can be appreciated in this instance, and Steve Robinson sees the main benefit as the increased flexibility in change of use, allowing café and restaurant uses to take up vacant retail units. This has introduced a degree of café culture back into the town, external tables and chairs being introduced with the only application required being a licence from the Highways Authority.

There was a 19.3% turnout for the referendum approving the NDO, in which 772 votes were made in favour of the proposals, compared to 496 against. The closer nature of this vote (61% in favour) compared to referendums on neighbourhood plans (average 89% in favour) probably reflects community fears about a 'planning free for all': the NDO is essentially a permissive rather than protectionist device.

There is no prescriptive approach put forward for neighbourhood planning, although good practice along with advice and funding is offered through the Department of Communities and Local Government (DCLG). Partnerships with non-governmental organisations such as Locality (capacity building), AECOM (technical support) and the mycommunityrights.org.uk website all seek to supplement the local support from local planning authorities.

Financial support and incentives for neighbourhood planning

Financial support has been provided to mobilise local initiatives, and a recent DCLG update[6] has stated that over £3 million in grants and £1 million in technical support has been awarded to local groups to support neighbourhood planning since April 2015. From April 2016, the grant available to an individual group involved in neighbourhood planning was increased from £8,000 to £9,000. Local planning authorities also benefit from funding which is triggered at each stage of the process, with payments made on neighbourhood areas being designated, neighbourhood forums being created, the submission of a neighbourhood plan and a final sum for each successful plan or order.

Since 2010, the government has also taken steps to incentivise local communities to allow or accept development. The new homes bonus was introduced in England by the Coalition government and remains in place today, although under review. The aim is to encourage local planning authorities to grant planning permissions for the building of new houses in return for additional revenue. Under the scheme, the government matches the council tax raised on each new home built for a period of six years. There is no hypothecation placed on any receipts.

Neighbourhood planning adopts a similar approach by rewarding local communities financially. Areas with a neighbourhood plan in place are able to receive up to 25% of the revenues generated from community infrastructure levy (CIL) associated with new development within their geographical boundary. The monies received can be used by parish or town councils to invest in local social infrastructure or community projects.

The Coalition government was explicit in its rationale for such incentivisation of new development: 'Instead of hectoring people and forcing development on communities, the Government believes we need to persuade communities that development is in everyone's interest. Incentives are key to getting the homes built we both need for today and for future generations'.[7]

However, there has been some scepticism expressed within Parliament about the actual impact of incentives in delivering new development. Scrutiny by the Public Accounts Committee and a subsequent report published in October 2013[8] found that 'The Department has yet to demonstrate that the new homes it is funding through this scheme are in areas of housing need and the Department's planned evaluation is now urgent.'

A subsequent evaluation published in December 2014 by DCLG[9] found that although there was evidence that the policy was beginning to impact positively on attitudes, such as support for new homes, the impact had been more limited in relation to plan making or planning decisions to date. Research associated with the evaluation revealed that out of 185 interviews conducted with local authority planning officers, 116 (58.9%) disagreed or strongly disagreed with the statement that, 'The New Homes Bonus has helped increase overall support for new homes within the local community', whilst only 20 (10.2%) agreed or strongly agreed.

Given that some research into neighbourhood planning suggests a distinct tilt towards protectionist policy agendas and greater adoption in more affluent areas (see Box 8.2 below), there has to be some doubt that financial incentives will have the desired impact of increasing a local community's acceptance or welcoming of new development. There also has to be a question as to whether local people either understand what financial benefits might accrue or whether they believe that they will ultimately benefit. However, plan making and delivery of development is a slow process, and it is probably too early to conclude whether incentive-based monetisation of neighbourhood planning is a model that works.

Box 8.2 Neighbourhood planning – popular yet protectionist and skewed towards southern affluent rural communities?

Neighbourhood planning has been embraced by many local communities throughout England, but it is clear from data provided by DCLG and others like Planning Resource that micro-level planning displays certain characteristics:

- By the end of 2015, some 126 referendums had been held;[10]
- Over 250,000 people have voted in neighbourhood planning referendums and all had been successful, with an average 89% vote in favour and an average 33% turnout;
- Some 1,700 communities representing some 8 million people are now engaged in neighbourhood planning; and
- In October 2014, the government[11] said that some £4.25 million had been awarded to community groups engaged in neighbourhood planning with a further £1 million expected to be allocated in the 2014/15 financial year. £22.5 million was budgeted for distribution between 2015 and 2018.

Research undertaken by Turley[12] found that of neighbourhood plans published up to February 2014:

- 91% were prepared by parish councils
- 55% had 'protectionist' agendas whilst 45% had 'pro-development' agendas;
- The average population of each plan area was 7,000

- The average score on the index of multiple deprivation for the area covered was 206 (the index ranges from 1 to 326, 1 being the most deprived)
- 73% of plans had been introduced within Conservative-controlled councils, only 9% were Labour-controlled
- 67% of plans covered rural areas and 33% urban areas
- 46% of plans have been published in the South East, 12% in the South West and 11% both in the East and West Midlands
- 75% of plans were from communities south of the line linking the Severn and Humber Estuaries

Data collected by Planning Resource[13] in May 2016 revealed:

- 1,947 applications for neighbourhood planning status had been made to local planning authorities of which 1,778 had been designated
- 450 neighbourhood plans were at pre-submission draft, 331 had been submitted, 230 independently examined, 192 had been the subject of a referendum and 151 had been adopted
- A correlation between engagement in neighbourhood planning and affluence – the top 30 most affluent LPAs (0.3% of the total number of LPAs) accounted for 17% of all referendums undertaken and 15% of all adopted neighbourhood plans. The comparative figures for the top 30 most deprived LPAs were respectively 0.2% and 2%.

From plan making to adoption

Once the neighbourhood area is designated, the plan-making process, led by the body that is taking it forward, should follow a familiar pattern of building an evidence base for policy and proposals; undertaking early initial community involvement; the drafting of the plan; pre-submission consultation; the submission to the local planning authority; the independent examination (including modifications); the holding of the referendum and then adoption following a positive vote in the referendum itself.

One of the key elements in the process is public consultation and involvement, and this is fairly critical given the 'Yes' vote that is required in the referendum if the neighbourhood plan is to be adopted.

Locality, in its Roadmap Guide,[14] seeks to set out good practice and underlines the fact that effective publicity and consultation is a statutory requirement, as well as ensures that in combination with the referendum 'democratic legitimacy' is achieved. Although not seeking to be prescriptive, Locality suggests a staged approach which involves the identification of key themes and issues through early involvement prior to the formulation of the plan's vision and aims; mid-stage involvement which seeks feedback on the draft vision and aims; and then finally consultation on the completed plan.

Clearly, the capacity and enthusiasm of the body taking forward the neighbourhood plan and the quality of advice and support it receives are crucial to the success of achieving local buy-in, but experience to date in the referendums has been a 100% success rate in the 126 referendums completed by the end of 2015 (see Box 8.2 above).

The robustness of local community involvement in neighbourhood planning is considered in the independent examination which forms part of the process. How this worked in one particular neighbourhood plan in the Oxfordshire town of Thame is examined in Box 8.3.

One of the frustrations evident among government legislators and officials has been the pace of neighbourhood planning in practice. There has been a recent impetus to simplify and expedite the process. This is reflected in the Housing and Planning Act 2016 which requires the application of time limits in relation to key stages in neighbourhood planning, such as the maximum number of weeks within which a local planning authority should designate a neighbourhood forum or hold a referendum once it has decided to do so.

The stages in a neighbourhood plan or order are summarised in Table 8.1.

Box 8.3 Case study: community involvement in the Thame neighbourhood plan

In his report to South Oxfordshire District Council, Independent Examiner Nigel McGurk scrutinised the public consultation undertaken on the Thame neighbourhood plan fairly closely, as there was some criticism from within the local community about one of the more advanced stages of the planning process.

This related to the identification of the preferred options for the plan, which resulted in the dispersal of some 700 new homes to six separate sites around the town. Some critics felt that this decision had been made 'behind closed doors' by a core group of town councillors and representatives of residents associations.

In considering whether the process was robust and fair, the examiner concluded that 'in order to progress the plan-making process efficiently, some decisions do need to be made by smaller groups. By their very nature, it is inevitable that smaller groups may not be fully representative of everybody with an interest.'

The examiner also concluded that the plan had been the subject of five stages of consultation and that this had exceeded the statutory requirements. Communication tactics deployed included:

- Newspaper advertisements
- Interviews on local radio stations
- Use of social media
- The erection of a large banner in a prominent location
- A website.

Two separate consultation weekends attracted 400 and 479 people respectively, and during the statutory six-week consultation period a series of five open workshops were held and attended by 85 members of the public. Prospective developers were invited to present their proposals and an exhibition took place at Thame's weekly market. Every household in the parish received a newsletter and there was a 'wrap around' edition of the *Thame Gazette* detailing key information. As a result, 246 responses were received on the draft plan.

There were attempts to try and engage with hard-to-reach groups. A focus group for young people was held, but there were few attendees, a problem often encountered by those involved in community involvement in planning.

In addition to the statutory consultation statement, Thame Town Council also produced a consultation report which summarised all the comments received at each stage of the planning process, with brief responses and references to specific changes to the plan which resulted.

The examiner concluded that an exemplary approach to neighbourhood planning had been followed.

Table 8.1 Stages in neighbourhood planning

Step 1	Identification and designation of a neighbourhood area (and a neighbourhood forum if required)	Local community identifies boundary for neighbourhood plan Local community applies to local planning authority for the area to be designated. Simultaneously the local community may apply for a neighbourhood forum to be designated if no parish or town council exists Local planning authority publicises and consults on the application(s) and makes a decision on the neighbourhood area

(Continued)

Table 8.1 (Continued)

Step 2	Initial evidence	Local community formulates vision and objectives, gathers evidence and drafts details of plan
		Local community consults on proposals for a minimum of six weeks
Step 3	Submission	Neighbourhood plan proposal and required documents are submitted to the local planning authority
		The local planning authority publicises the proposal for a minimum of six weeks and invites representations
		The local planning authority arranges an independent examination of the neighbourhood plan or order
Step 4	Examination	An independent examiner makes a recommendation to the local planning authority on whether the draft neighbourhood plan or order meets certain legal tests
		The local planning authority considers the report and decides whether the neighbourhood plan should proceed to a referendum
Step 5	Referendum and enactment	A referendum is held
		If a majority support the neighbourhood plan the authority must bring it into force

Neighbourhood planning – impact and analysis

There can be little doubt that the promotion of micro planning has been taken forward enthusiastically since it was first introduced after the change in government in 2010. There is also every sign that the momentum will continue.

A paradigm shift in plan making has occurred and the contrast between the previous Labour government's regionally based spatial planning and current practice is dramatic, although as noted above the delivery of neighbourhood planning has not been universally applied. Given that research suggests that around 90% of neighbourhood plans have been taken forward by parish or town councils, it is possible to extrapolate a neighbourhood planning 'penetration' of about 20%, based on the number of applications made for neighbourhood planning status.

Given the fact that there are around 9,000 councils at this lowest (or 'first tier') level covering a population of around 16 million (25% of the population in England), there is clearly much more scope to introduce micro planning. However, it is evident that even with a much higher level penetration, the vast majority of the English population is unlikely to participate. That said, there is the possibility of creating the bodies to transact neighbourhood planning in urban areas; the Kentish Town neighbourhood forum in the London Borough of Camden is but one example of this.

The experience of neighbourhood planning to date has also been by its very nature highly varied in terms of the size and character of neighbourhood areas; the legislation tends to be very flexible in this regard. This raises questions as to whether there could be a lack of spatial and policy cohesion.

It has been proven that where neighbourhood planning has been introduced, it appears to have been popular. Many people have been engaged and involved in a process which might be characterised by some as previously the closed domain of elected members or local government officials (although community involvement programmes have always been undertaken by local planning authorities – a process which continues to operate alongside neighbourhood planning). There is an empirical case for neighbourhood planning having 'popularised' or even 'democratised' planning (although see below for an examination of the issues around the use of referendums).

It might be interesting to carry out a piece of research to discover the extent and success of local community involvement in areas before and after neighbourhood planning was introduced. One way of measuring this might be to look at this purely in the context of participation in referendums, but there is no 'control' comparator here. The results of community participation are also quantified in the evidence base around neighbourhood plans in terms of people attending events and submitting views. Such data is also available for district or borough plan making, so a useful comparison could be made.

The representativeness or otherwise of local opinion in a community is always a keen subject of discussion in relation to planning and development. There is never one single or uniform expression of community opinion. Reactions to development are multi-faceted and nuanced, and often opinions are influenced by proximity to proposed development or even the stage in which an individual or group becomes aware of or involved in the process. This is often a key frustration in community involvement when early, front-loaded consultation is organised and plans or policies shaped accordingly, only to then be undone by late, often last-minute objectors who have not been involved previously. Campaign or lobbying groups seeking to influence outcomes on planning often direct most of their energy at the final stages of decision-making.

A key issue is whether the transition to more locally based or parish-based decision-making has actually resulted in greater acceptance of development or has led to increased housing delivery.

The evidence is inconclusive, although an analysis by DCLG[15] of a very small sample of the first 16 neighbourhood plans revealed that these plans had delivered a 10% increase above and beyond that provided by the extant local plan and that planning applications and permissions had been advanced rapidly.

The thesis that neighbourhood planning has delivered more development than would have otherwise been achieved through traditional LPA-wide plan making has however been questioned by Turley's research, which suggests a bias towards protectionist rather than pro-development agendas.

Certainly, the experience of developers including Bloor Homes[16] over recent years has been that the toughest tier of English government in trying to secure support for new development has been at parish or town council level, although investment in relationship building and a willingness to engage with local communities has paid dividends in villages in East Anglia, where the company has promoted development.

On the other hand, the company has encountered strong resistance in other village communities where even very modest levels of development have been presented by objectors as potentially causing wide-ranging damage to the local social fabric. In this respect, although the basis on which the company engaged was broadly similar, the community reaction was diverse and depended on the unique circumstances of each locality.

Turley's research tends to support the wariness of the development industry towards neighbourhood planning and suggests a fear that micro-level planning is intended to set or amend strategic policy rather than elaborate upon it, increasing risk and uncertainty. This in some instances has led to a decision by developers to bring forward proposals in advance of neighbourhood planning to try and limit any restrictive policies that might be put in place.

An additional issue relates to the willingness of those involved in neighbourhood planning in embracing developers as stakeholders in the process, rather than private organisations whose primary motivation is to generate financial returns from land. In this respect, the body leading the neighbourhood planning exercise may simply consider the views of commercial organisations as irrelevant or self-interested, when in fact the experience of developers and their perspectives can actually be a positive aid in bringing forward sustainable development and sound local planning policy. As noted above, the attitude of plan makers on this point is likely to be directly related to whether they have a permissive or restrictive stance towards growth and development in their area.

The referendum-based approach to neighbourhood planning also raises a number of important questions. Often in controversial planning cases, the narrative put forward by objectors centres on an argument that it is 'not wanted' by the local community. This is not surprising given that few people encountering the planning system have a detailed understanding of the way it works.

A prime example here has been NewRiver Retail's experience of promoting small convenience stores. In Dudley Metropolitan Borough Council, objectors were mobilised based on arguments that there are already existing convenience stores nearby, even though the planning system is clear that competition amongst such stores is not a material planning consideration. Observation of the Dudley MBC Development Control Committee in 2015 reveals that an assessment of local opinion (usually measured by numbers of objections or the views of a local ward member) has been very influential in members of that committee rejecting professional officers' advice.

The referendum-based nature of neighbourhood planning perhaps reinforces the feeling that planning is in fact a 'democratic exercise', and whether something should go forward should be based on whether local people want it. Furthermore, given that the referendum is a 'take it or leave it' vote on what is likely to be a complex matrix of policy issues, there is a degree of doubt as to how informed the electorate might be about the issues at stake. Clearly, this can be addressed by robust and extensive communication with the local community so that an informed decision can be made.

It is possible to argue that a referendum-based approach to neighbourhood planning cannot be fully inclusive or actually tackle the key issues at stake. For example, should votes be given to 16- or 17-year-olds? Should all employees in an area be given votes? In the absence of the special arrangements for business areas, should businesses have a vote on all neighbourhood plans? Should families with children have more votes given their potential investment in the future? Should the votes of those immediately adjacent to proposed development sites be given more weight, or maybe less? Should there be separate votes on different chapters of plans? Should there be votes on competing plans?

This, perhaps, returns to where we started on neighbourhood planning – the tension between micro and macro policy objectives, particularly given the presumption in favour of sustainable development contained in the National Planning Policy Framework and the stated desire to increase the supply of housing to meet wider social and economic needs.

Some important questions arise out of this tension. For example, where local and neighbourhood plans are being progressed at the same time, should the neighbourhood plan give more weight to the existing plan or to the emerging plan? How can a neighbourhood plan, in general conformity with an out-of-date local plan, meet the needs of the community going forward? Should a local plan be able to override a neighbourhood plan once it has measured its objectively assessed need (OAN) for new housing, if more homes are needed?

Inevitably this has already played itself out through the courts, where conflicts have emerged between policies set out at neighbourhood level and national policies aimed at delivering housing (see Box 8.4).

Box 8.4 Case study: the frontline of conflict between local and national planning

Sayers Common – Woodcock Holdings, Mid-Sussex

The then Secretary of State Eric Pickles overturned a planning inspector's decision in September 2014 to approve a 120-home scheme by Woodcock Holdings in Mid-Sussex. The inspector had supported the appeal after finding that the local authority could not demonstrate compliance with a key national planning policy – the requirement for a five-year supply of housing land.

Reversing the inspector's decision, Pickles's decision letter said that the emerging Hurstpierpoint and Sayers Common 2031 neighbourhood plan, having been submitted for examination in May, could now be given more weight. The neighbourhood plan identified housing allocations elsewhere and concluded that the Woodcock scheme was therefore in conflict with the neighbourhood plan and premature. Pickles did concede that the view of the inspector was correct in finding that the Woodcock scheme was compliant with the National Planning Policy Framework.

In the High Court in February 2015, the judge ruled in favour of Woodcock's challenge of the secretary of state's decision finding amongst other things that:

- Emerging plan policies must be considered 'out of date' where there is no five-year supply of housing judged against objectively assessed housing needs. Emerging neighbourhood plans were not in any way immune from this requirement and sit lower in the policy hierarchy than do national policies
- Clear reasons are needed for giving more weight to the neighbourhood plan process than to evidence-based housing needs which have been identified at a district-wide level
- The neighbourhood plan adoption test is far more limited than for local plans. The neighbourhood plan process may not even be considering important planning considerations such as whether enough land has been allocated to meet housing needs.

This is likely to have implications for those pursuing a more 'protectionist' agenda at neighbourhood level, as it suggests that only where OAN is fully met in a neighbourhood plan will it be considered to be sound. Only those seeking to meet needs in full or over-allocated sites (not evident in any great respect in neighbourhood planning to date) could be seen to be compliant.

Although seeking to empower local communities, the complexity of plan making may result ultimately in a feeling of disempowerment, as those promoting development resort to legal challenges based on errors made by inexperienced local groups and people. It is also not possible for neighbourhood planning to address the same issues that local plans need to take into account.

This has certainly been the result of the recent legal challenge by developer Lightwood Strategic to the Haddenham neighbourhood plan (Aylesbury Vale District Council), where a judicial review was launched due to alleged mistakes on the preparation of the proposed site allocations for new development. In March 2016, the district council decided not to contest the legal action following advice from its counsel, and this resulted in the removal of the housing policies in the adopted Haddenham neighbourhood plan.

The expectations created by Localism may have been too great, and the limits perhaps are now beginning to be more greatly appreciated given the dynamic nature of plan making and the hierarchy of policy which applies. Certainly, the statement of the Haddenham Parish Council reveals disappointment:

> Neighbourhood Planning is now likely to turn into a costly process that is carried out by professionals for those communities with sufficient funding and will be beyond the reach of smaller communities. This surely goes against the principles of Localism on which the Neighbourhood Planning policy was founded.[17]

Perhaps this is a fairly accurate recognition of the realities of the system. What set out as a simple political objective of giving power to local communities has found itself embroiled in an inherently complex environment. Neighbourhood planning is certainly here to stay, and it will be interesting to see how politicians propose both to extend and strengthen the existing rights and to correct anomalies in the system.

Notes

1 Conservative Party (2010) The Conservative Manifesto: Invitation to Join the Government of Britain. London: The Conservative Party
2 Ibid.
3 See in particular the Neighbourhood Planning (General) Regulations 2012. London: HMSO
4 The Stationery Office Limited (2016) *The Queen's Speech.* London: HMSO
5 The Stationery Office Limited (2011) *Localism Act* Section 61H to Schedule 9. London: HMSO
6 DCLG (Department for Communities and Local Government) (2015) *Notes on Neighbourhood Planning.* London: DCLG
7 DCLG (Department for Communities and Local Government) (2013) *Communities to Receive Cash Boost for Choosing Development* [Online]. Available www.gov.uk/government/news/communities-to-receive-cash-boost-for-choosing-development [Accessed 12 October 2016]
8 The Stationery Office Limited (2014) *Report of the Committee of Public Accounts: The New Homes Bonus* [Online]. Available www.publications.parliament.uk/pa/cm201314/cmselect/cmpubacc/114/114.pdf [Accessed 12 October 2016]
9 DCLG (Department for Communities and Local Government) (2014) *Evaluation of the New Homes Bonus* [Online]. Available www.gov.uk/government/publications/evaluation-of-the-new-homes-bonus [Accessed 12 October 2016]

10 DCLG (Department for Communities and Local Government) (2015) *Notes on Neighbourhood Planning*. London: DCLG.
11 DCLG (Department for Communities and Local Government) (2014) *£23 Million to Get More Neighbourhood Plans Across England* [Online]. Available www.gov.uk/government/news/23-million-to-get-more-neighbourhood-plans-across-england [Accessed 12 October 2016]
12 Turley, H. (2014) Neighbourhood Planning. Plan and Deliver? London: Turley.
13 Planning Resource (2016) *Map: Neighbourhood Plan Applications* [Online]. Available www.planningresource.co.uk/article/1212813/map-neighbourhood-plan-applications [Accessed 12 October 2016]
14 Chetwyn, D. (2016) Neighbourhood Plans, Roadmap Guide Locality. London: Locality.
15 DCLG (Department for Communities and Local Government) (2015) *Neighbourhood Planning: Progress on Housing Delivery* [Online]. Available http://my community.org.uk/wp-content/uploads/2016/08/Neighbourhood-planning_-progress-on-housing-delivery-.pdf [Accessed 12 October 2016]
16 Based on community engagement work carried out on the company's behalf by Polity Communications Ltd
17 Haddenham Parish Council (2016) [Statement in relation to the application for a judicial review of the Haddenham neighbourhood plan].

9 Localism and new community rights

Introduction

Localism does not begin and end with neighbourhood planning. The 2011 Act and secondary legislation which followed put in place a number of other community-focused initiatives which have met with varying degrees of success. Community Right to Build, Community Right to Bid, Community Right to Challenge, and other 'rights' were introduced, as their names suggest, to increase the role of *community* in planning. Not only do they require substantial levels of involvement on proposed changes, but the changes themselves are instigated by and for the community; local people remain in control throughout the process, and fundamentally it is the community, rather than the local authority, which gives permission for the change to go ahead. A form of delegated powers and driven by a commitment to devolve power to communities, this might, in principle, be considered just one rung off the top of Arnstein's Ladder of Citizen Participation.[1]

The initiatives also sought to tackle local issues such as the housing shortage, neighbourhood blight caused by empty buildings and vacant land, and the demise of traditional community infrastructure such as village pubs.

Community Right to Build

Community Right to Build allows communities to propose and develop housing schemes, community facilities, places of work or other land uses. Planning consent is granted through an external examination resulting in a development consent order. This removes the requirement for the development proposal to go through the standard planning process provided certain criteria are met and procedures carried out, including a referendum of those living in the local area.

The Community Right to Build process potentially provides substantial opportunities for community involvement:

- To **establish community support**, the individual or group must hold discussions with the community and stakeholders such as the local

Table 9.1 The Community Right to Build

Step 1	Establishing community support	The leading individual or group must consult local residents to determine levels of support for the project and establish a support base.
Step 2	Creating a legal entity	The project must be put forward by a parish council, neighbourhood forum or community group which has been established with the purpose of 'furthering the social, economic and environmental well-being of individuals living, or wanting to live, in a particular area'.
Step 3	Defining the neighbourhood area	The sponsoring group must make an application to the local authority to confirm the geographic boundaries of the site.
Step 4	Developing the business case	This significant stage requires that the wider community is consulted, and that a comprehensive plan is put in place which is financially viable and identifies other partners (such as private developers, housing associations and the local authority), if any. Grants are available to cover funding of this stage.
Step 5	Preparing a Community Right to Build Order	The Community Right to Build Order must be drawn up and made available to the wider community and statutory consultees[2] for their comments.
Step 6	Submitting a Community Right to Build Order	The Community Right to Build Order is then submitted to the local planning authority with accompanying documents.
Step 7	Independent examination	The local authority arranges for an independent examination to take place. The examiner recommends whether the application should proceed to a referendum, with or without amendments.
Step 8	Local authority approval	The local authority determines whether it is satisfied that the order meets the basic conditions.
Step 9	Referendum	The local authority organises a referendum, which is open to all within the defined area. It may be broadened to include others who will be impacted by the future development.
Step 10	Result	If the referendum achieves 50% support or higher, the Community Right to Build Order is granted.

authority, planners and funders. A strategy to ensure that local people are involved includes presentations to associations, interest groups, schools and community councils. Public meetings may be held to explain the purpose of the Community Right to Build and what it can achieve and provide updates on a regular basis.

- One of the group's first jobs is to form a **legal entity**. This gives it credibility and establishes good governance, may enable charitable status to be granted and identifies roles and responsibilities.

- **Defining the neighbourhood area** can sometimes involve two parishes working together or perhaps a parish council working with a neighbouring industrial estate or town centre forum.
- **Developing the business case** takes place over time and can have many re-iterations and re-writes because typically projects and ideas appear and then are modified or dropped. Typically this starts with a search for the asset required (land) or negotiations for an existing asset. Viability of the scheme is the determining factor – what will the asset cost, and can the organisation afford both to acquire it and undertake the transformations required? Land values can change and negotiation is always a tricky undertaking.
- **Preparing and submitting a Community Right to Build Order** usually requires some professional expertise, and to obtain this the organisation will often have to raise funds to pay for professional advice.
- The **independent examination** may require representation from the local community, something which would need to be coordinated.
- Following **local authority approval**, the **referendum** is the final stage of the process, and perhaps the most time-intensive. Significant time and funds will be invested in communicating the benefits of the Community Right to Build, promoting the referendum and motivating the local community to vote.

In the first five years of the legislation, only three Community Right to Build Orders had been granted, all to one parish council – a significant disappointment and embarrassment for the proponents of Localism.

Community Right to Build is mostly being pursued in the case of community facilities, where some funding is accessible to local groups. Community housing – which had achieved success many years before the advent of Localism – continues to flourish in many areas, but in most cases community groups have chosen to follow the traditional planning route rather than the new and more complex process of the Community Right to Build. Those few Community Right to Build Orders which have been granted have come about in conjunction with a neighbourhood plan which is required to follow a similar format.

Box 9.1 Case study: the first Community Right to Build scheme

Ferring Parish Council – West Sussex

The village of Ferring was required to allocate land for 50 new homes. A site, already in the ownership of a major housebuilder, had been thought suitable by Arun District Council, but local people thought otherwise: their preference was to spread the new homes around the village, to introduce the new housing in phases and, specifically, to provide homes for downsizers. Ferring Parish Council took advantage of the new Localism initiatives by first drafting a neighbourhood plan which specified the location of the new homes and then, by using Community Right to Build, to provide appropriate housing.

In developing the plans, Ferring Parish Council, working with external consultants, identified land suitable for housing. This included allotments which were deemed too small and the plot occupied by the ageing village hall. By allocating housing on these sites, not only was the parish council able to deliver the required homes, but also allocate new land for both facilities.

As the land in question was already in community ownership, the Community Right to Build process was a fairly straightforward one. Public consultation, from December 2013, highlighted the need for more detailed plans to be drawn up for the three proposed sites. The consultants, working with a committee of residents and three charities affected by the proposals, applied for a Housing and Communities Agency (HCA) grant in in early 2014 and drew up plans for the proposed housing and village hall.

In addition to ongoing discussions within the community, a six-week consultation was carried out, culminating in a referendum in December 2014, at a cost of £75,000.

The referendum supported the Community Right to Build, but as Ferring District Council has yet to acquire the freehold for all of the land involved, there is some way to go before the plans are fully realised. The parish council was frustrated with public sector involvement – specifically the lack of any precedent or framework at a district, county and national level, and felt that the process was extremely cumbersome and would have been prohibitively expensive had it not been possible to obtain grants from the HCA and Locality.

Box 9.2 The Community Right to Build vanguards

Comment by Laurence Castle, project officer at South Cambridgeshire District Council

South Cambridgeshire District Council was one of the 11 Community Right to Build Laurence vanguard councils chosen by the housing minister in September 2014 to pioneer the Community Right to Build scheme. South Cambs has been given a grant of £50,000 to bring forward a minimum of 100 plots of land for custom builders. We currently have 285 people on the register and are progressing a pipeline of sites for sale, as well as exploring joint ventures and other ways of bringing self- and custom-build plots to the market.

My role as the project officer is to encourage and facilitate take-up for the Community Right to Build and assist self-builders in achieving a development consent order. This has been achieved through connecting future self-builders with land; advising them of useful organisations such as the Community Land Trust, running workshops on issues such as legalities and technical skills; providing links to funding organisations (a funding mechanism is being put in place with Capita which provides early access to finance); providing a wide range of information and, importantly, sharing best practice across the vanguard councils and more widely.

Take up of the Community Right to Build has been slow so far, but we have a very encouraging pipeline of projects and the systems in place to support them and for a sub-regional roll out with partners which is being discussed.

Community Right to Challenge

Like the Community Right to Build, the Community Right to Challenge occupies a high position on the Ladder of Citizen Participation. A means by which local authority management can be replaced, it enables community groups to take over the running of public services. Like the Community Right to Build, however, its take-up has been slow, suggesting little appetite for citizen control.

Through the Community Right to Challenge, voluntary and community groups, charities, social enterprises, parish councils, or two or more individual members of a local authority may submit an expression of interest to run a service which then triggers a competitive tendering process.

The initiative is driven by a recognition that many people love, and will go to great lengths to protect, their local services, perhaps also influenced by local authorities' financial pressures and their need to maximise profit on disposals.

The Community Right to Challenge can be applied to almost all local services, including planning departments. Table 9.2 describes the process.

Anecdotal evidence from those who have been involved in promoting the right suggests that there has been some improvement in the dialogue between commissioners and local groups, but with the exception of a single reference to the right in case law[3] there has been no actual use of the right.

Table 9.2 The Community Right to Challenge

Step 1	The community group identifies the service and the potential improvements that can be made
Step 2	The community group ensures that structures, skills and resources are available
Step 3	The community group submits an expression of interest to the local authority
Step 4	The local authority considers the expression of interest
Step 5	The expression of interest accepted or rejected (specific grounds are necessary for rejection)
Step 6	A procurement exercise is carried out
Step 7	If successful, the community group takes over the running of the service

Table 9.3 The Community Right to Bid

Step 1	A community forum, neighbourhood forum or parish council identifies and nominates the asset as having 'community value'
Step 2	The local authority lists the asset, having consulted with the owner
Step 3	The asset is added to the local authority's database, which it has a duty to publicise and maintain
Step 4	When the owner announces their intention to sell the asset, a six-month moratorium is put in place, allowing the community group to assemble a business plan and make funding arrangements. The community group is likely to benefit by contacting the owner of the asset to discuss either a possibility of bringing the project forward in partnership or agreeing sales terms.
Step 5	At the end of the moratorium period, the owner is free to sell to whomever they wish, at the price they wish. This may be by private arrangement, through the local authority or at auction.

Community Right to Bid

The Community Right to Bid is a less adversarial means by which a local organisation may seek to preserve buildings of particular relevance to the community. Received more enthusiastically than other rights, the Community Right to Bid allows communities and parish councils to nominate land or buildings to be listed by the local authority as an 'asset of community value' if its principal use furthers the community's social interests or has the ability to do so in the future. Village shops, pubs and libraries frequently fall into this category. At the point at which the asset is put up for sale, a moratorium of up to six months allows the community the opportunity to raise the funds to make the purchase.

Although the fundraising stage sometimes puts an end to communities' ambitions, there are numerous success stories.

Box 9.3 Case study: Community Right to Bid

The Ivy House pub – Nunhead, South London

When, in April 2012, the Ivy House pub was put on the market and its tenant landlord given five days' notice in which to leave, local residents became very concerned as to its future, but found themselves unable to prevent the sale due to the considerable price tag and the fact that the transaction took place quickly and smoothly.

However, when the pub was put back on the market in October 2012, the residents acted quickly. Although unable to purchase the pub at the asking price of £750,000, they were able to utilise the Community Right to Bid to have the facility listed as an Asset of Community Value. In doing so, they gained a six-month moratorium.

During this time, they formed a team which included a property lawyer, a conservation adviser and a town planner and set about mobilising residents and lobbying politicians. Social media, a website, leaflets, posters and public meetings were all deployed, and local networks including Peckham Vision, the Peckham Society and Friends of Nunhead Cemetery were utilized. The group ensured that the pub was listed by English Heritage, thus limiting physical changes to the building.

Thanks to a conditional loan finance from the Architectural Heritage Fund and a grant through the My Community Rights programme, the community group bought the freehold of the pub, which re-opened in August 2013.

The pub is already providing a variety of community uses: a large ballroom area has been brought back into use as a venue for live music, theatre and comedy and hosts children's events and dance/yoga classes during the day, and a smaller back refectory area is used as a meeting space for many local groups and societies on an ad hoc basis. Business support and legal advice from the Plunkett Foundation and the sale of community shares in the pub have helped to raise the funds necessary to carry out the work.

Following the success of the project, the group is organising a skills bank to enable local organisations to learn from its expertise.

As the example of the Ivy House pub shows, an indirect impact of the Community Right to Bid is that residential developers with plans to convert pubs and schools into housing have frequently been frustrated by the moratorium, which not only creates delay and uncertainty but can result in the local community campaigning to preserve the original use of the building.

Community Asset Transfer

The Community Asset Transfer is similar to the Community Right to Bid in that buildings or land may become the property of a community group if it serves, or has the potential to serve, a community-related function. Unlike Community Right to Bid, however, Community Asset Transfer involves the transfer of ownership at less than market value.

The Community Asset Transfer is technically a policy rather than a right and as such can be arrived at in a variety of ways.

Of all the products of Localism, it has perhaps the most potential to unite communities in a common purpose and revitalise a crumbling asset. Previously failing schools, court buildings made redundant by geographic reorganisation and churches affected by dwindling congregations have all benefited from the initiative.

Box 9.4 Case study: Community Asset Transfer

Bramley Baths – Leeds

Bramley Baths, located on the outskirts of Leeds, is an Edwardian facility comprising swimming baths, a gymnasium and Russian steam room.

In September 2011, the city council proposed to reduce the facility's opening hours from 90 to 48 a week. The reduced hours and threat of closure was a great disappointment to the local community and prompted voluntary agency BARCA to establish both a 10-person steering group and larger organisation of supporters, the Friends of Bramley Baths, to keep the facility open.

A series of public meeting were organised and at an early stage, shortly after the advent of Localism in June 2011, it was decided that Friends of Bramley Baths would take the baths into community ownership via a Community Asset Transfer.

Initially the group submitted an expression of interest (EOI), which was accepted in August. The EOI outlined that the Friends of Bramley Baths would run the baths for 49 hours a week, rising to 60.

Eight local engagement events were held and the Friends of Bramley Baths carried out substantial research into local residents' likely future use of the facility. A finance team carried out detailed research into budgeting and human resource issues to ensure the project's viability.

On the basis of a positive response to the research and a promise of funding from Keyfund, a social investment financier, Leeds City Council was supportive of the application and approved the Community Asset Transfer on 1 January 2013.

Three years on, the Baths are now open 105 hours per week, employ 25 members of staff, teach children from 14 local primary schools and provide private swimming lessons for over 900 local children. The Baths have begun to generate a small annual surplus and have a bright future.

Community Right to Contest

The Community Right to Contest is a policy variation on the Public Request to Order Disposal (PROD) rules which were put in place in the Local Government and Planning Act 1988, in that it allows community bodies to request the release of land in public ownership providing a convincing case can be made for its future use. Unlike the PROD rules, however, it covers not only disused and redundant land and buildings and but also that which is currently occupied and could be put to better use.

There is little, if any, evidence of the Community Right to Contest having been used during the first five years of Localism.

Conclusion

The success of Localism initiatives varies considerably. Certainly in the case of Community Right to Build, there is a general sense that the procedural complexities outweigh the advantages. In the cases where the initiatives have been deployed more widely, there were similar mechanisms in place prior to Localism.

So has Localism, though these initiatives, met its objectives? The debate probably needs to begin with a definition of the word 'community' and address whether the very word can be something of a euphemism. In reality, the 'community' which steers any of these initiatives is typically a small group of informed and educated individuals – those capable of understanding planning law and with the requisite time, inclination and confidence. An entire geographic community is far too large and diverse to be jointly responsible for a development initiative and in reality the right can only be taken up when a community organisation acts on behalf of a wider community. And although this wider community has the option to become involved in the decision-making by voting in a referendum, it could be argued that this 'tick-box' exercise facilitates participation no higher than the fourth rung of the Ladder of Citizen Participation. Furthermore, despite having its roots in the locality, a community group may not be as well equipped to run a comprehensive consultation as a professional organisation – for example, having the abilities and resources to target hard-to-reach groups and to process and analyse results – and consequently the level of involvement will be less effective.

Another frequent criticism is that this is a site-specific, piecemeal and thereby non-strategic approach to development: although there is a need to comply with existing neighbourhood plans and local plans, community-led planning is rarely initiated in the light of these more strategic documents. And despite these being intended as community-led initiatives, sceptics have argued strongly against the powers being bestowed to the community, suggesting for example that the Community Right to Challenge process can be used as a 'Trojan horse' enabling private developers to take over public services by the back door.

Whether initiatives such as these succeed in uniting a community are also debatable. In many cases, both parties admit that the process has done little to strengthen relationships between local authorities and their residents, something that many would regard as an important foundation for a successful local community. The Community Right to Challenge, in particular, can theoretically disempower local authorities and result in acrimony between individuals, local authorities and the private sector. In the context of the Community Right to Bid, too, developers have stated that although community assets are rarely sold to the community as the result of the right, the very process of identifying 'valuable' community assets can significantly stifle future development. Several instances of derelict pubs being ear-marked for development and their sale complicated by the right have been cited in the planning media.

To capitalise upon the positive facets of Localism, local authorities should seek to ensure a clear channel of communication (perhaps through a Localism 'champion'), respond positively to requests for information, engage thoroughly with all local partners and provide community organisations with support in the form of advice, expertise and access to funding where possible. As the case studies have demonstrated, Localism can result in positive change given the right set of circumstances, and it is in local authorities' interests to work with communities to achieve this.

Despite additional community rights having been suggested (including most notably a Community Right to Beauty which proposes measures to give communities powers and incentives to shape, enhance and create places of aesthetic merit) and a possibility of extending the moratorium period in the Community Right to Bid, it would appear that the appetite for Localism has been minimal. Therefore, it is unlikely that Community Right to Build, Community Right to Challenge and Community Right to Bid and the other policy initiatives will significantly change the way in which communities interact with developers and local authorities in the future.

Notes

1 Arnstein's Ladder of Citizen Participation was described in more detail in the introduction to this book.
2 The local authority will advise on the specific bodies to be consulted. These may include the county council and parish council; voluntary bodies whose activities

fall within the defined area; business groups; bodies which represent the interests of specific groups such including faith, race and disability groups in the defined area; the owner of any of the land within the defined areal, and English Heritage. Depending on how the proposal may affect them, the following may also be determined as statutory consultees: the Coal Authority, the Homes and Communities Agency, Natural England, the Environment Agency, Network Rail Infrastructure Limited, railway operators, the Highways Agency, the Marine Management Organisation, the health trust, the electricity provider, the gas provider, the water and sewerage companies, the Garden History Society, the Civil Aviation Authority, the Secretary of State for Defence, the Secretary of State DEFRA, the Secretary of State Transport, the mayor of London, British Waterways Board, Sport England, the National Park Authority, the Health & Safety Executive, local highway authority, the Theatres Trust.

3 Draper v Lincolnshire County Council 2014 (concerning the closure of libraries in Lincolnshire).

10 The planning application process

The functioning of the planning system as we have examined elsewhere varies in England, Scotland, Wales and Northern Ireland, but its common thread is that it is plan-led and largely administered by local government in terms of plan making, determining of planning applications and taking enforcement action against unauthorised development.

In England, district councils are responsible for the majority of planning matters except transport, minerals and waste planning, which are usually the responsibility of county councils. In some areas of the country, there are single-tier authorities which combine both district- and county-level planning responsibilities. In London, there are strategic planning powers, where planning applications of a certain scale or significance may be determined by the London mayor.

As can been seen in Box 10.1, the number of planning applications administered by local authorities in England is substantial and requires considerable human and financial resources to be deployed. Different local authorities experience different pressures and needs, and it is not unusual for many – especially in areas where there is a high cost of living – to experience staff recruitment difficulties.

The vast majority of planning applications in England are in fact decided by professional officers through delegated authority rather than going through a formal decision-making process involving elected members. Recent reforms of the planning system relating to permitted development rights also mean that certain types of development, such as domestic extensions or changes of use, do not require planning permission (although they may be subject to a prior approval process).

There are a number of different types of planning applications, but broadly the classic division is between applications for full planning permission and applications for outline permission. However, the variants include applications for reserved matters, discharge of conditions, amendments to existing permissions and certificates of lawful development. All of these types of application can require elements of community involvement, depending on the specific circumstances faced.

Box 10.1 Key statistics on planning applications in England, 2016[1]

In the first quarter of 2016, English district-level planning authorities:

- Received 119,700 applications for planning permission, down 1% on the corresponding quarter of 2015
- Made 98,400 decisions of which 91,700 (93%) were delegated to officers
- Granted 86,200 decisions, up 3% from the same quarter in 2015; this is equivalent to 88% of decisions, up one percentage point on the same quarter of 2015
- Granted 11,300 residential applications, down 1% on a year earlier.

Overall, 9,000 applications were for prior approval for permitted development and 7,400 of these (82%) were approved without the need to go through the full planning process.

In the 12 months ending June 2016, the same planning authorities:

- Granted 378,200 decisions, up 4% from the figure for the year ending June 2015
- Granted 47,600 decisions on residential developments: 6,000 for major developments and 41,600 for minor.

The key stages in the planning application process

From a community as well as a political involvement perspective, the easiest way to look at the planning application process is in terms of three stages:

- Pre-application
- Submission and predetermination
- Determination

At each of these stages, various activities and interventions need to be considered in terms of community involvement and dialogue with key political and civic stakeholders. We now consider each of these stages and the practical considerations which apply.

Box 10.2 How a typical local authority sees the application process

The usual advice given by local authorities tends to focus on the transactional nature of the planning application process. The London

Borough of Brent is typical in this respect in the way that it describes the process from the council's perspective. It explains it as having the following six steps:

Step 1 – Pre-application advice

Before you decide whether to make a planning application or not, we highly recommend that you obtain pre-application planning advice from us.

We will be able to advise you whether your proposal is likely to be approved or not, and can recommend changes to ensure that your planning application has the best chance of success.

Step 2 – Application and validation

We strongly recommend that you apply online through the Planning Portal, where you will be advised of the documents you will need to submit and the correct fees, which means that your application is more likely to be validated.

Once submitted, applications are checked to ensure all documents and fees required are correct.

Any missing information will be requested before processing can start. We normally aim to process your application within eight weeks of receiving a valid planning application, so submitting the correct information the first time can help avoid delays.

The timescale for major planning applications is 13 weeks.

Step 3 – Consultation and publicity

Consultation letters are sent to neighbours and, where applicable, various bodies to obtain their expert view.

Advertisements, where required, are placed in the appropriate local paper and on site.

Others can view plans and comments can be made online.

The consultations period is 21 days from the date of publishing.

Step 4 – Site visit and assessment

The site is inspected and the application assessed by the planning case officer, taking into account planning policies, consultation responses and public representations.

Where relevant, the planning officer will also gather any site-specific information such as photographs.

Step 5 – Recommendation

The planning officer will make a recommendation, via the 'officers' report', on the application to the person or body authorised to make a decision.

Case officers do not make the final decision on applications. The officer's report will include all of the relevant facts relating to the application in order to inform the decision maker.

Step 6 – Decision

A decision is taken on the application by the appropriate body.

With most householder applications, senior officers who have delegated authority from the planning committee normally make the decision under what's known as 'delegated powers'.

This means that they can make the decision without going to the relevant committee, which speeds the process up.

Around 95% of householder applications are decided this way.

Applications can be approved or refused.

Approved applications will often have conditions attached that must be complied with.

If an application is refused and the applicant does not believe the decision was correct, they have the right to appeal.

Pre-application

This is the most important stage to get right given the emphasis on the frontloading of consultation on planning applications. The ability to demonstrate that those interested in the final form of a planning application have had a genuine opportunity to influence it before it has been resolved is very important.

The public and civic groups can be very sceptical about whether their views will be taken on board and will often express the view that the process is a 'box-ticking' exercise rather than a genuine attempt to listen and respond to community feedback.

The critical element here is ensuring that the process starts early enough and certainly some considerable time before a planning application is submitted. Carrying out local community involvement and submitting an application a week or two later is not going to build confidence.

Successful community involvement exercises involve building relationships and also establishing trust. There is significant value in seeking to understand who the key local stakeholders are and then actively seeking them out before commencing any wider community involvement. Local councillors should be the first port of call and can be an excellent source for understanding who are the most important leaders or influencers within a local community – after all, they are likely to be encountering them all the time – and there is every good reason to identify and approach them for advice at an early opportunity.

Other good sources in terms of understanding the community environment are the local media, local websites, social media and recent planning applications that have been through the decision-making system. Representations made on nearby planning applications can indicate who will need to be engaged by applicants. This also will reveal the statutory and non-statutory consultees that typically intervene in a particular local planning authority.

Box 10.3 Active engagement with civic groups at the pre-application stage

Kensington & Chelsea

Brockton Capital and U+I decided to take forward comprehensive proposals for Newcombe House, a 1960s office block at a key location in Notting Hill Gate within the Royal Borough of Kensington and Chelsea. They took the decision from the outset to identify key civic groups within the locality and actively involve them on their emerging plans for the regeneration of the site.

Dialogue with key local residents' and stakeholder groups began in 2013, when separate briefing/feedback workshops of around 2–3 hours were organised with representatives (officers or post holders) of the Kensington Society, the Cherry Trees Residents' Amenities Association, the Pembridge Association and the Notting Hill Gate Improvements Group.

These were the principal civic groups with an interest in the area at the time either because of their geographical proximity to the site or because of their involvement in planning and regeneration matters within the Royal Borough.

This early engagement involved a presentation to the residents and stakeholder groups by the architects, Urban Sense Consultant Architects (USCA) that focused on:

- Key elements of the site analysis
- How the site analysis shaped the urban design strategy
- Proposals for new perimeter buildings and public space
- Design development for a potential 'Corner Building' to be sited near to the Notting Hill Gate/Kensington Church Street junction.

At this point in the design process, the Corner Building had a horizontal slipped form, and feedback was received on this concept along with the proposed approach with the public realm and perimeter buildings.

Following the decision of the design team to take a new approach with the form of the Corner Building in response to comments from both the local authority and local community, a new series of workshops were set up in September 2014 and these took place with the same groups.

As previously, the format for the sessions was the same for each group and involved a presentation by the architects USCA, which focused on:

- A recap of the key elements of the site analysis
- Revised proposals for the perimeter buildings and public space
- Revised proposals for the Corner Building.

For the first time in the community involvement process for the scheme, indicative computer-generated visuals of the emerging scheme were presented.

The design team then undertook a further period of design review and development and then re-engaged again in early 2015 with the residents' and stakeholder groups. A new workshop was organised with the groups and for the first time, a session was organised with representatives of the Hillgate Village Residents' Group.

The applicant was not previously aware of the existence of this group, which had been brought to its attention following a meeting with the local ward councillors, and made an approach to meet with them as soon as it had been provided with the relevant contact details. The group comprised residents living in the closest proximity to the development site. In itself, this demonstrated that it is always possible that new stakeholder groups can emerge during the course of a pre-application process, particularly if it is a lengthy process around a complex and significant scheme.

The session with the Kensington Society was an update on progress with the scheme and also an opportunity to re-examine key issues that had been raised previously. The session with the Hillgate residents largely followed the approach taken with the initial briefings of the various groups and allowed the design team to explain the rationale and content of the emerging proposals for the first time to this audience, as well as receive general and more specific feedback from residents most closely neighbouring the western boundary of the Newcombe House site.

In addition to the workshop sessions, public exhibitions were also organised both during the pre-application stage and at the point of submission of the planning application. The intended submission scheme was also presented to the stakeholder groups.

Throughout the process, which involved some 17 separate workshop sessions, the design team gathered and responded to the feedback received. This resulted in specific changes to the scheme, such as a decision not to proceed with any vehicular access (except emergency vehicles) to the proposed new public square. It also led to the inclusion of new health facilities within the proposed development to address community needs and desires.

The views of the different groups were by no means the same, and each had specific concerns and interests. This revealed an important aspect of community involvement – the community is often diverse and certainly rarely speaks with one voice. Proximity of a resident's property to a site raises very different issues than a more holistic desire to see long overdue regeneration come forward. This was ultimately demonstrated by the support given by the Kensington Society to the planning application and objections by the Hillgate Village Residents' Group.

In formulating a community involvement programme, applicants need to review each local authority's Statement of Community Involvement, as this will set out expectations in terms of the precise methods of community involvement which should be deployed.

It is also good practice to discuss with officers the intended programme that will be undertaken, as different authorities often take varying approaches. With major or complex development proposals, or where there is a significant interest from the local authority's point of view in terms of land ownership, dialogue with senior elected members may also be advisable. Clearly, meeting officer and political expectations in community involvement is an important factor in ensuring that there is confidence in the process of taking forward a planning application.

As can be seen in Box 10.4, there are a number of tactical interventions that can be deployed as part and parcel of an integrated pre-application community involvement programme. Key elements which should be considered include:

- Initial involvement with politicians and officers.
- Early identification and contact with key stakeholder groups – these can be resident groups, business groups, special interest groups and even national pressure groups.

- A clear view of the principal community targets for any involvement – this is not a precise science, in that deciding, for example, on a catchment for the delivery invitation letter to a community involvement event is always going to be based on a degree of subjectivity. The problem here is that what the planning system might view as an impact and therefore the degree to which nearby residents should be contacted may well be different to the view of the community or elected representatives.
- The style of community involvement that is required – in some instances, a more intensive form of community involvement such as Enquiry by Design (EbD) should be contemplated, where the local community is invited to take a much more inclusive role in the development and design of major proposals. This is often used in the case of new settlements or large strategic developments.
- The number and location of any community involvement events are also critical. It is important to provide accessible locations at times when the majority of the local community can attend. This might include the need for weekend sessions or multi-venue activities on schemes which may have major coverage, such as large town centre regeneration proposals.
- The need to arrive at a coherent combination of traditional and online methods of involvement. Different demographic and age groups respond differently to face-to-face contact such as at exhibitions and online contact via websites or social media. A mixture needs to be considered to reach different audiences.

Although applicants cannot be required to undertake pre-application community involvement, failure to do so, even when an application does not generate significant objections, can create a negative environment at the eventual decision-making stage. As we have examined elsewhere in this book, the growing culture of community involvement and the emphasis on Localism means that is not possible to advance planning applications of any scale without having a robust and acceptable programme of pre-application community involvement.

Although there can be no guarantees that community involvement will lead to a planning application being supported by a local authority, investment in a thought-through programme will undoubtedly reduce risk, allow the opportunity to remove elements which could lead to refusal, and also identify individuals or groups in the community which may well lend their support.

A well-presented Statement of Community Involvement as part of a planning application submission which details activities undertaken, sets out quantitative and qualitative results, as well as demonstrates clearly that proposals have been revised or refined in direct response to community feedback can be significant in shaping a local authority's response to a planning application and ultimately on final decision-making.

Box 10.4 Case study: a typical local authority SCI

Middlesbrough Council

In formulating a community involvement programme, the Statement of Community Involvement of the local authority in which a development proposal is located needs to be reviewed and understood at an early stage. In practice, many local authorities adopt similar approaches, but it is important to understand any idiosyncrasies and also discuss with officers and in some cases elected members.

Middlesbrough Council's Statement of Community Involvement (adopted March 2016) states:

The onus of pre-application publicity rests with the developer. The council will expect developers to undertake pre-application community engagement in the following instances:

- Major developments[2]
- Environmental impact assessment applications which are accompanied by an environmental statement
- Proposals which depart from (i.e. do not conform with) the development plan
- Proposals which affect a public right of way/footpath
- Any development proposals which the local planning authority determines as having significant planning policy implications
- Any development which appears to be significant and/or controversial.

Generally, the level of community engagement will be tailored to reflect the scale and likely impact of the new development. Developers are encouraged to discuss their engagement ideas with the council as part of their standard pre-application discussions. Where a development is anticipated to be controversial, advice will be offered.

As a guide, (depending upon the scale and impact of the proposal) developers will be expected to:

- Notify local residents by letter of the proposed development, and where they can get further information
- Put out a press release detailing their proposed development (and where people can get further information) in cases concerning land or a building to which the public has enjoyed access or been part of public life
- Put together letters and/or press releases which give details of a public engagement event that the developer has organised

- Arrange a public engagement event that details the developer's aspirations for the site, the scope for public comment, how comments will be dealt with, and what happens next
- Arrange the public engagement event to take place at a time which enables attendance from as wide a cross-section of the public as possible
- Arrange the engagement event to take place at the site/building to which the proposal relates, or the nearest publicly accessible meeting venue (the council will provide a list of suitable venues upon request)
- Engage with local community/councils and interest groups within the surrounding area of the proposed development as part of their pre-application community engagement procedures.

Where a developer has carried out pre-application community engagement, as a matter of good practice, they should include a statement with their application which details what has been done – scale of notification, numbers attending the event, comments made or received, copies of the applicants' response to the comments received and any resultant revisions to the scheme.

Where significant changes have been made to the prospective planning application, and depending on the extent and nature of these changes, the council may advise the developer to consult further, pre-submission.

Government policy is that planning authorities should not refuse to accept an otherwise valid application because the applicant has failed to carry out pre-application community engagement. The failure, however, by a developer to undertake community engagement before submitting their proposal could result in avoidable objections being made which could be material to the determination of the application.

Submission and predetermination

Once a planning application has been submitted or is on the verge of being submitted, there is often a good case to employ community involvement and communications tactics to signal the conclusion of the pre-application process.

This could involve dialogue with stakeholder groups that have been previously involved; use of print and broadcast media; briefings of politicians; updating of websites with planning application documentation; and the holding of public exhibitions or events where the application proposals can be presented along with any changes that may have resulted from the pre-application community involvement.

Although sometimes seen as a luxury, such activities at the point of submission or just after can generate goodwill and demonstrate transparency.

Box 10.5 Case study: community involvement at submission stage

NewRiver Retail – Burgess Hill

NewRiver Retail developed extensive proposals to redevelop Burgess Hill town centre and had undertaken a programme of pre-application community involvement, where the public was invited to participate in community exhibitions of the emerging plans to create new homes, a new hotel, a new multiscreen cinema, restaurants, a new library and modern retail units aimed at leading high street names.

During the pre-application process, a proposal had been presented to the public which involved the relocation of the Lidl supermarket in the town centre to a site close to the town's railway station. This proposal was required to facilitate the wider town centre regeneration but had generated some opposition from local residents nearby to the relocation site. As a result, NewRiver worked with Lidl to examine alternatives.

A new relocation site was identified in a different part of Burgess Hill at a former gasworks site. The submission exhibition which was held enabled NewRiver to demonstrate that it had responded to feedback that it had received earlier in the process. The event also gave an opportunity to display the latest computer-generated images as well as an animated fly-through of the proposals. The company also worked closely with the Burgess Hill Town Council.

A separate community involvement event was then organised for the new relocation site where residents were invited to attend and give their feedback.

During the period between the submission of the planning application and its eventual determination, the monitoring of representations to the local authority is an important task for applicants. Most local authorities today have the ability to look at representations online, and the content and source of objections or expressions of support are important to understand. It is likely that members involved in the planning decision will become aware of such representations or indeed be contacted directly by the individuals or organisations that are making them.

Despite the best efforts of applicants in seeking to build consensus around planning applications and also after having spent significant time in involving the community, it is often the case that objectors await the final stages in the life of a planning application to launch campaigns of objection. It is inevitable that those opposed to planning applications will focus their energy on decision makers at the final stage, and it is important for applicants to take steps to provide appropriate responses as required.

Applicants themselves may of course actually encourage those groups and individuals they may have encountered during pre-application community involvement to submit supportive representations. Although the number of representations is not a material planning consideration, it has been shown time after time that if there are significant numbers of objections without countervailing expressions of support, then this could have a significant impact upon politicians during decision-making. The extent to which this applies varies across local authorities and also at different times in the electoral cycle.

The reality is that objectors are often more motivated, with supporters taking a more passive stance or not wishing to reveal their identities in a situation where they may be subject to criticism from other members of the local community. It is where strong relationships have been built during pre-application community involvement programmes that stand the best chance of delivering supportive representations.

The key unknown is always whether local authority planning officers are going to recommend the application for refusal or approval. Long observation of planning decision-making reveals that the prospects of overturning a recommendation for refusal at committee are far less than the eventuality of elected members overturning a recommendation for approval.

Sometimes, depending on the relationship with the local authority's officers, it is possible to get a fairly clear early steer on the recommendation they will make to the committee considering a planning application. An understanding of what is likely to happen can allow applicants to either take defensive or offensive action. One such example of further community involvement activity post submission but prior to determination is described in Box 10.6.

Box 10.6 Case study: post submission opinion polling

Next plc – Plymouth

As part of a rollout of its new generation multi-format stores incorporating fashion, homewares and a garden centre, Next plc selected a site in Plymouth adjacent to an existing retail park for a new store.

During the course of protracted negotiations and discussions with officers of Plymouth City Council, it became apparent that officers were not prepared to recommend approval to city councillors.

Next had undertaken extensive community involvement work which had included a two-site exhibition and use of social media. The response from the public had been enthusiastic. 302 out of 313 feedback forms received had revealed support, and a dedicated Facebook page 'Next4Plymouth' had received over 2,700 'likes'. In addition, some 470 individual representations had been made to the city council, with 97% indicating support.

Despite the significant support that local community involvement work had revealed, Next decided in the lead-up to planning decision-making to commission a Market Research Society–accredited company to undertake some independent and representative opinion survey work to test the views of local people further.

One thousand Plymothians were randomly contacted across a sample of Plymouth postcodes and were asked whether or not they welcomed the new store. Some 94.7% of those contacted indicated that they did welcome the new store.

The result of this research was then given to the local media in the run-up to the city council planning committee and was featured prominently. Elected members were also sent a briefing document which summarised the results of the survey and also highlighted the acceptability of the new store in planning terms.

Despite the recommendation of officers for refusal which was explained in detail to members at committee, a majority of the councillors decided not to accept their officers' advice and approve the application. They concluded that the range of planning benefits outweighed the concerns expressed by officers on retail impact on the city centre.

One other possibility, which appears to be increasingly offered by local authorities, is the opportunity for applicants to make a separate presentation to elected members prior to determination. Some local planning authorities try to make this happen during the pre-application process, although it is often the case that it is done post submission. In both the London Borough of Hounslow and the Royal Borough of Kensington and Chelsea, such presentations are open to the public who are allowed to ask questions or make comments.

Given the limited time at committee and also the often limited time of members to fully understand complex planning applications, such opportunities to provide presentations should be grasped by applicants. Often, key issues or concerns that members have are revealed, and if necessary action can be taken or information provided that can address any concerns, potential problems at the point of decision-making can be avoided.

Determination

After many months, or often years, of work and the submission of an expensive planning application which may require a complex suite of reports and technical information, it usually comes down to a vote of elected members at a local authority committee, unless an application is dealt with under delegated powers by officers.

In the lead-up to the committee itself, the officer's report is published and sometimes is not available until close to the committee itself. The minimum period for disclosure in England is four working days, excluding the day of the committee meeting, meaning that there may be limited time to make any further responses.

Applicants often need to consider whether they should make direct representations or contact with members involved in decision-making. It is very often the case that objectors are doing this, and well-organised campaign groups can mobilise emails or other contacts aimed at members, with contact details freely available on local authority websites.

One tactic that can be used is to provide a briefing document, which is sent to members just prior to committee after the officer's report has been analysed. This document can summarise the key planning issues and also relay the benefits in a way which is likely to impact positively with politicians. One of the approaches taken by Bloor Homes with its planning applications is to not only stress the planning merits but also underline and quantify the socio-economic benefits of their proposals. The company has used graphics to illustrate direct and indirect job creation, the amount of new homes bonus that might apply, the amount of council tax that could be generated by new residents occupying the new homes that they construct and also the consumer spending power which the residents could bring to a local area.

At the point of decision-making, many local authorities also undertake committee site visits. These vary in their nature, and it is important to understand how these operate in individual authorities. Sometimes there is the opportunity for applicants or their representatives to attend and deal with specific questions which members might have prior to committee. However, this is not always the case, as some authorities do not allow applicants or other interested parties to be present. There have been instances where protesters attend such site visits, but in one recent example[3] in the London Borough of Ealing, the aggressive nature of the protest by objectors had the counterproductive effect on politicians who felt that they were being unfairly and unduly pressurised.

The final opportunity for applicants to put their case is at the committee itself. Most local authorities now allow for applicants and objectors to committees prior to a decision. It is important to check the rules which apply here, as often prior notification deadlines and other rules apply. Time limits are enforced on speeches – three to five minutes is usually the norm – and different authorities have different approaches in relation to elected members asking questions of those individuals or groups which address them. Cross-examination can be sometimes protracted and detailed.

It is generally advisable to prepare a speech, although ensuring that this is delivered in an engaging way is as important as the content. Although given the multidisciplinary nature of most teams that work all major development proposals, it often is much better if a single individual delivers a speech and that other team members are on hand as necessary to answer specific questions which relate to their particular expertise.

Although there are many similarities between the approach and procedures followed by individual committees involved in planning decision-making, it is important to understand that every local authority has its particular style and culture. Time attending planning committees to observe their dynamics is time well spent, as intelligence gained through such observation can be helpful in terms of any future interventions that need to be made.

Conclusion

This chapter has focused on the planning application process and the practical implications for community involvement at the various stages of the process, from pre-application to determination at committee. There are a number of considerations which apply and tactics which need to be deployed depending on the individual circumstances.

No one planning application or local planning authority is the same, and there is no 'one size fits all' approach that should be employed. In the next chapter, we examine the interface between officers and elected members and the political context in which planning operates.

Notes

1 Department for Communities and Local Government, Planning Statistical Release, June & September 2016
2 Major developments as defined in the Town and Country Planning (Development Management Procedure) (England) Order 2015
3 'Residents fear prejudice over Gypsy Corner planning decision' see www.getwestlondon.co.uk/news/west-london-news/residents-fear-prejudice-over-gypsy-9375699

11 The role of local authorities in considering and determining planning applications

Roles and responsibilities of officers and elected members

Necessarily, in fulfilling planning functions and ultimately determining planning applications, there are distinct roles and responsibilities for a local authority's officers and its elected members. Both officers and members are public servants, but the former is responsible to the local authority as a whole, while the latter owes ultimate responsibility to the local electorate.

While there is a clear difference in roles and responsibilities between officers and members, it is vital for the operation of the planning system that there is a successful working relationship between the two. This often involves sensitive handling and mutual understanding of the respective roles.

It is often possible to discern the balance of power between officers and members within different local authorities. Some councils tend to have an internal culture whereby there is more of an emphasis on political direction and intervention; in others there can be more of an 'officer-led' tendency. Planning decision-making can illuminate this balance, revealing whether officers take a more passive role at committee or even in terms of whether officer advice tends to be accepted.

Officers are employed by local authorities to provide impartial professional advice in relation to planning matters. They are the principal point of contact with applicants and their advisers and provide the consistent interface in various stages of planning decision-making. In terms of planning applications, they often fulfil a critical role in assisting applicants in framing policy compliant proposals.

With the current emphasis on pre-application discussions – aimed at ironing out any policy conflicts and fashioning proposals which will be acceptable to a local authority – this advisory and often negotiation-based interaction with applicants has become increasingly important in larger development proposals and is often extensive.

However, the new approach in the 2011 Localism Act on issues around predetermination (see below) has meant that there is now more space for elected members to be involved in pre-application discussions. Ironically, this can cause objectors to become suspicious that officers or members may be assisting applicants 'behind closed doors'. Ultimately, when applications are decided in planning committee by elected members, the officers' report to a committee or subcommittee will recommend approval or refusal, but it is entirely in the hands of members to decide whether to implement this advice.

The potential for conflict in the member/officer relationship flows primarily from a tension between the two roles which elected members perform. On the one hand, they are required to transact planning decisions by using planning criteria, but on the other, they act as representatives of public opinion in their local communities. The balance of this dual role varies considerably among individual members and indeed in different local authorities. It is often a critical factor as to whether members accept or reject their professional officers' advice.

The respective roles of officers and members was considered by the Third Report of the Committee on Standards in Public Life, which was chaired by Lord Nolan and published in 1997.

In a dedicated chapter of the report on planning, it stated:

> It is essential for operation of the planning system that local concerns are adequately ventilated. The most effective and suitable way this can be done is via the local elected representatives, the councillors themselves. Councillors owe their position to the electorate, and decisions they take they are, by and large, attempting to serve the community which elected them. More specifically, they cannot put out of their minds the plain fact that in order to be re-elected they have to secure sufficient votes from the electorate.

The report added:

> Councillors themselves may be influenced by feelings which do not derive from dispassionate examination of the planning issues. They may see themselves as leaders of local opinion rather than judges, and they may even have been elected on a specific platform of opposing or supporting a particular development or type of development. In our view, if planning decisions by local authorities were to be regarded as quasi-legal decisions, in which councils played a role similar to that of inquiry inspectors or judges, there would be no point in involving councillors in such decisions. They might as well be taken by planning officers, or by inspectors.[1]

Planning decision-making is often made amid controversy, and development proposals or proposed local plan allocations result in significant public interest and comment. Lord Nolan recognised that planning is probably the most contentious matter with which local government deals. The reason is that fundamentally the planning system is seeking to reconcile private and public interests. The scope for conflict was summarised very well by the Planning Advisory Service in a guide produced for the Local Government Association:

> One of the key aims of the planning system is to balance private interests in the development of land against the wider public interest. In performing this role, planning necessarily effects property interests, particularly the financial value of land holdings and the quality of their settings. Opposing views are often strongly held by those involved.[2]

Recent reforms of the planning system in England have stressed the notion of a presumption in favour of sustainable development and furthermore the National Planning Policy Framework states:

> Local planning authorities should look for solutions rather than problems, and decision-takers at every level should seek to approve applications for sustainable development where possible. Local planning authorities should work proactively with applicants to secure developments that improve the economic, social and environmental conditions of the area.[3]

This has an important bearing on decision-making, and it is often the case that objectors to development struggle to understand why there is a positive presumption of this kind when latent or organised major local opposition exists to proposed development. Objectors naturally develop a narrative that development is either 'not needed' or 'not wanted' regardless of its planning merits, and it follows as a result that elected members should support the community's 'view'.

The key point here is that planning decision makers are obliged to only take into account material planning considerations,[4] and although public views can legitimately form part of elected members' deliberations, those views must relate to relevant planning matters. In practice, and in particular where members of the public or opposition groups have mobilised significant opposition, the need to focus exclusively on material conditions sometimes becomes secondary or marginalised. This can be brought into sharper relief if there are wider political sensitivities such as those which revolve around forthcoming elections (see Box 11.2).

It is difficult for those members of the public who have limited knowledge of or exposure to the operation of the planning system to understand that mere support or opposition is not sufficient reason to grant or refuse planning permission. It is equally difficult for elected members who are reliant on votes from their local electorates to divorce themselves from feedback from the community.

Community feedback may be based primarily on the activities of organised campaign groups using various methods, such as objection letters, petitions and interventions in traditional and social media, to exert pressure on politicians making planning decisions. Such community feedback may be highly localised or skewed towards certain socio-economic and age groups. Support for development may result from mobilisation of sections of the community that may benefit as end users.

It is often very difficult for elected members to take a clear view on whether the feedback received through statutory consultation on planning applications is indeed representative of the community as a whole and may even involve trying to second guess whether there is a 'silent majority'. More experienced members or those not operating in sensitive or marginal political environments tend to be less influenced by community-based campaigns, as the balance they strike is often towards the planning merits rather than the level of opposition.

Recommendations are more often than not presented by officers as an 'on balance' decision. The material planning considerations are weighed up, non-material considerations are eliminated and a conclusion is drawn. It follows that elected members may see the balance differently and come to a contrary conclusion – this is the inherent nature of planning decision-making and also explains why outcomes are often difficult to predict. It is not an entirely objective process as a proposal's economic development benefits, for example, may be viewed differently by individual members. This is why, as we will explore further below, it is useful to have an understanding of individual members' attitudes or positions on previous applications.

Box 11.1 Case study: political sensitivity in decision-making

Essential Living – Swiss Cottage, London

Leading private rented sector (PRS) developer and operator Essential Living acquired a key site at Swiss Cottage in the London Borough of Camden. The site was at a highly sustainable town centre location, immediately above the London Underground station, and Camden Council's planning policy identified the site as being suitable for a tall building.

Essential Living consulted locally on a mixed-use scheme that included residential, commercial and community uses, but it became apparent very early on that there was a fundamental disagreement between local residents and local civic groups about the height of the proposed building, which at 24 storeys they considered out of scale with the nearest properties.

There was little or no opportunity to build any consensus around the future of the prime regeneration site, and the focus of leading local campaigners was to generate as many objections as possible. This was achieved through petitions, door knocking the local area, use of the media (the *Camden New Journal* and the *Ham & High*), public meetings, use of social media and ultimately turning up in substantial numbers to Camden's planning committee.

Although the local community around the site was opposed in principle, other key consultees such as the London mayor and the design council were supportive of the proposals. This led to Camden's officers recommending approval of the application.

Although the planning case for the development appeared to be strong (this ultimately being proven by a successful appeal), the political environment faced by Essential Living was negative. The planning application was seeking to navigate the decision-making process at a particularly sensitive time in the lead-up to both the borough elections and the general election in 2014.

The ward in which the site was located was at the time a split Conservative/Liberal Democrat ward within the Labour-controlled borough, and the parliamentary constituency was one of the most marginal in the country, having been held by Labour in the 2010 general election by just 42 votes out of a total of 52,822 votes cast. Moreover, the new Labour parliamentary candidate, Tulip Siddiq, was seeking to make her mark and retain the seat. She took a stance against the development.

The decision came before the planning committee in September 2014, but the success of the local campaign and the pressure it was able to exert in a politically marginal area, combined with time sensitivity in the lead-up to important elections, was a major factor in influencing the decision of councillors to overturn their officers' recommendation to approve the scheme.

Politics continued to play itself out even after the refusal, as Conservative London mayoral candidate Zac Goldsmith raised the issue as part of his campaign at the commencement of the public inquiry in 2015. Tulip Siddiq, now the new MP, also continued to campaign on the matter, even raising it at Prime Minister's Questions in the immediate aftermath of the secretary of state's decision to accept an independent planning inspector's decision that the proposal should be allowed to proceed.

General conduct of elected members and officers

One of the key outputs of the Committee on Standards in Public Life in its First Report (1995) was to propose 'Seven Principles of Public Life', which were described in Chapter 2. These have been adapted over the years, and the Localism Act 2011 makes it a requirement that local authority codes of conduct are consistent with the principles. Furthermore, since August 2012, all local authorities have been required to produce local codes of conduct.

Related to the codes of conduct are registers of members' interests and the requirement to disclose pecuniary interests. Many local authorities also produce associated codes of practice in planning which supplement the overarching code of conduct and the council's standing orders.

Local authority staff who are chartered town planners are bound by the Royal Town Planning Institute's Code of Professional Conduct,[5] which has five core principles relating to competence, honesty and integrity; independent professional judgement; due care and diligence; equality and respect; and professional behaviour. Most local authorities also have adopted codes of conduct for employees, and there is also statutory legislation[6] which places restrictions on senior officers holding membership of political parties or elected office in another council.

Box 11.2 Plymouth City Council's planning committee code of practice

In September 2013, Plymouth City Council produced a comprehensive document called *Probity in Planning*,[7] which is part of the council's formal constitution. The objective of the document is to adapt national guidelines on probity to take account of the local situation in Plymouth.

The code covers the basis on which both elected members and officers should determine planning applications submitted to the city council and covers key issues, such as:

- Drawing a distinction between members' separate roles on the planning committee and their representative role as a ward member – 'the Planning committee must not favour any person, group or locality'
- Passing on lobbying correspondence received by members to officers
- Avoiding any exercise of pressure by members on officers for a particular recommendation
- Indicating any instances of excessive lobbying or inappropriate offers of gifts or hospitality
- The need for complete impartiality by officers and their compliance with guidance from appropriate professional bodies
- The prohibition of using political group meetings to dictate how members should vote on planning applications ('whipping') or the lobbying of fellow members

- Explaining the difference between predetermination, where a member has made up his or her mind on voting a certain way prior to a committee meeting, and predisposition, where a member can take an initial view about a planning matter yet maintain an open mind
- Identifying instances where planning committee members dealing with contentious applications in their own wards may withdraw from voting
- How being a recipient of lobbying from interested parties is acceptable as long as members retain an open mind and do not express an intention to vote one way or the other before a planning committee takes place
- The basis on which ward members can receive informal briefings from applicants at pre-application stage
- The requirement that any meetings between planning committee members and applicants should always involve a planning officer
- The base-level training requirements that members must complete in order to fulfil their planning decision-making functions.

Registration and declaration of interests

A key issue for elected members generally and particularly in relation to the determination of planning applications is the registration and declaration of interests.

Councillors and officers are required under statutory legislation and regulations to register disclosable pecuniary interests. It is a criminal offence to fail to do so within 28 days of election, or a change in circumstances or by participating in a meeting which considers a matter where a councillor or officer has such an interest.

There are broadly three types of interest: personal, prejudicial and pecuniary. It is the responsibility of individual members to seek advice from each local authority's monitoring officer and declare such interests.

Plymouth City Council's *Probity in Planning* document also sets out the basis on which the declaration and registration of interests should work. The city council states that

> Members should also not give grounds for a suspicion that any such interests may arise and have not been declared. When an application is to be determined, there is an expectation that it will be dealt with transparently, openly and in a fair way. Members will be expected to take account of all relevant evidence and give it appropriate weight in the decision making process and arrive at a reasoned sound decision.

Box 11.3　Declaring and registering interests

Plymouth City Council's code of practice

There are three types of interest, personal, prejudicial and pecuniary. A member will have a personal interest in a planning committee decision if:

- If the matter relates to an interest in respect of which the member has given notice in the statutory register of members' interests; or
- The decision might reasonably be regarded as affecting their well-being or financial position or that of a relative or friend or employer to a greater extent than other council taxpayers, rate-payers or inhabitants of the authority's area.

Where a member considers he or she has such a personal interest in a matter, he must always declare it. A personal interest will become a prejudicial interest if a member of the public with knowledge of the relevant facts would reasonably regard that interest as so significant that it is likely to prejudice the member's judgement of the public interest.

Where any member of the council considers that they have a prejudicial interest, they should discuss this situation with the monitoring officer or his representative as soon as they realise this is the case:

- Complete a standard form and pass it to the democratic support officer prior to the start of planning committee
- Ensure they do not participate at any stage in the consideration of the planning application if following advice from the monitoring officer they consider that any decision they take could be challenged on the grounds of bias
- Ensure they do not seek or accept any preferential treatment, or place themselves in a position that could lead to the public to think they are receiving preferential treatment, because of their position as a councillor.

With regard to planning officer interests, the assistant director for planning services will check the officer declarations list and advise the officer accordingly if they feel that alternative arrangements are necessary in presenting the report.

Pecuniary interests are defined in regulations. The regulations in general will mean that a member will have to register any interest they or their husband or wife or civil partner has in:

- Any employment, office, trade, profession or vocation that they carry on for profit or gain

- Any sponsorship that the councillor receives including contributions to the councillor's expenses as a councillor; or the councillor's election expenses from a trade union
- Any land licence or tenancy they have in Plymouth
- Any current contracts leases or tenancies between the council and them
- Any current contracts, leases or tenancies between the council and any organisation with land in Plymouth in which they are or have a partner, a paid director, or a relevant interest in its shares and securities
- Any organisation which has land or a place of business in Plymouth and in which they have a relevant interest in its shares or its securities.

If a councillor has a pecuniary interest, they must leave the meeting immediately and cannot participate, or participate further, in any discussion of the matter at the meeting, or participate in any vote, or further vote, taken on the matter at the meeting.

Sometimes there are grey areas around whether members should or should not participate in planning committee meetings. An example was at a meeting of the Dudley Metropolitan Borough planning committee, where a member had neglected to update his declaration of interest. An application was before the committee to convert a public house to a convenience store to be operated by the Co-operative Food. The member concerned had previously been a member of the Cooperative Party, the political wing of the cooperative movement. Although these two arms of the cooperative's organisation are totally unrelated and have no direct relationship, the member concerned withdrew from the meeting and did not participate.

Such decisions are always matters for individual members, although in the case above it is questionable as to whether there was any prejudicial or pecuniary interest involved. The matter was controversial locally, and it is possible that the member concerned did not want to be involved in decision-making. Is not unusual that councillors – particularly those that represent wards in which applications are located – withdraw from meetings on the basis that they have an interest.

Applicants, supporters or objectors in planning decision-making are able to inspect registers of members' interests: these are usually found on individual local authority websites and can often provide information about members which may be relevant to key planning issues which arise. A checklist at the end of this chapter covers issues which may help in understanding the context of decision-making in planning and assist with how applicants, supporters or objectors may present their respective cases through political advocacy.

Predetermination and predisposition

One of the more confusing areas of the 2011 Localism Act relates to its Section 25. This developed a new concept in relation to planning decision-making, whereby elected members should not be regarded as having a 'closed mind' on planning applications if they make a public statement or are involved in some public activity which could indicate (directly or indirectly) their likely view on a planning application.

The distinction is drawn between members clearly expressing an intention to vote in a certain way prior to committee meeting (predetermination) and taking or expressing a view on a planning application but maintaining open mind.

In practice, for many involved in navigating the planning system, it is quite difficult to conceive of a situation where an elected member involved in planning could take or express a public view – given that this in itself is often in response to members' assessment of public opinion or the merits of a proposal – and at the same time retain an open mind.

It is only when planning committee members express fundamental sentiments against certain types of development and indicate that they will always oppose such development that predetermination becomes clear and could be the subject of judicial challenge.

It is the nature of politics for politicians to take stances on certain issues, and this often occurs early on in a process. Given the representative role of elected members which we examined earlier in this chapter, is often very difficult for members to credibly change their position when a decision has to be made. This calls into question to some extent whether it is really possible to have an 'open mind' after a public stance has been taken on a planning issue.

Lobbying of elected members

In political life generally increasing focus has been placed on issues around lobbying. There have been a number of moves to try and regulate lobbyists and those engaged in seeking to influence political decision makers.

In its report on lobbying,[8] the Committee on Standards in Public Life stated:

> We reaffirm that lobbying is a legitimate and potentially beneficial activity. Finding opportunities for individuals and organisations to talk to policy and decision makers and legislators is part of the process by which policy is formulated, implemented and tested. Free and open access to government is necessary for a functioning democracy as those

who might be affected by decisions need the opportunity to present their case.

However, the very term 'lobbying' has negative connotations and is often presented as being underhand, secretive or improper. As former Prime Minister David Cameron said in 2010:

> [S]ecret corporate lobbying, like the expenses scandal, goes to the heart of why people are so fed up with politics. It arouses people's worst fears and suspicions about how our political system works, with money buying power, power fishing for money, and a cosy club at the top making decisions in their own interest.[9]

Within the arena of planning decision-making in local government, lobbying by interested parties is regular and extensive. However, although undoubtedly applicants routinely engage in activities which seek to shape and influence planning decisions, the reality is that the vast majority of lobbying received by elected councillors originates with objectors; it can also often involve councillors lobbying their fellow councillors. As we have set out in Chapter 10, this plays itself out regularly at planning committees when members take advantage of speaking rights to address their colleagues.

A typical example was a planning application taken forward by shopping centre owner NewRiver Retail in Worthing, Sussex, in June 2016. The application involved the loss of some mature trees, and the lobbying campaign which took place on this was led by the Worthing Society. They employed the use of a street stall to gather a 3,000-name petition, made interventions in the local media, directly contacted elected members and ultimately attended a planning committee meeting to press their case.

Today, applicants have to assume that civic groups and individual members of the public are using a variety of techniques to lobby elected members. Digital communication channels such as emails and social media have increased the opportunities to reach politicians. Given this environment, it is critical that applicants or promoters of development take a proactive stance to present their side of the story or risk refusal. Not intervening in the debate and failing to rebut objections or ensure that wider benefits are fully understood can often be fatal for the prospects of individual planning applications.

Given that lobbying of elected members making planning decisions is so widespread, applicants should positively engage with councillors to ensure that they are aware of all of the relevant issues. For example, in the case of the proposed investment in Worthing town centre by NewRiver, the economic impact on the local economy of not allowing the proposed application needed to be conveyed to councillors to ensure they had the full picture. This was vital in achieving the eventual support of the planning committee.

Planning applications often become focused on individual issues to the detriment of wider implications or issues. It is the role of lobbying to ensure that this does not lead to decisions based on incomplete or partial information. Carried out professionally and transparently, and with regard to relevant codes of conduct, such political advocacy is a vital element of good decision-making and the operation of local democracy.

Understanding and responding to the political environment

Operating effectively within the planning system demands that the representative role of elected members alongside their planning decision-making role is fully appreciated, and those involved in navigating the planning system need to develop clear strategies and tactics.

A checklist of some of the issues and tactics that need to be taken into account by applicants and those involved in seeking to influence local plan making is set out in Box 11.4.

Box 11.4 Environmental considerations

A political and community audit is often an important precursor to embarking on development proposals or interventions in local plan making. This can examine a number of issues, such as:

- The political stability of the local authority – does it often experience changes in political control or does it tend to be controlled by one political party? Do controlling political groups tend to have a stable political leadership or are there frequent changes?
- The regularity of elections – many local authorities have annual elections, which tends to make decision-making particularly sensitive in the three- to four-month period prior to May when elections usually take place
- Is the local authority member-led or officer-led? What is the nature of working relationships between officers and members? Has the membership of the relevant committee or committees been consistent or changed regularly over recent years?
- How independently minded is the committee which deals with planning applications – does it generally follow officer recommendations?
- How authoritative or respected is the chair of a committee and how long has he/she been in place?
- Is a development site, or a site which is being promoted within local plan process, located in a politically sensitive area – whether

by virtue of the fact that it is marginal in electoral terms or because it is represented by senior politicians within the authority?

- Is there strong political leadership within the local authority in terms of economic regeneration, and what role do senior politicians play in this?
- What is the size of the committee which determines planning applications on the plan allocations? What is the composition of this and what experience do individual members have? What interests have been declared by members? Where do members live in relation to proposed development sites? Is there anything from past decision-making which suggests that members have certain preferences or attitudes towards development?
- What do the council's standing orders or constitution reveal about the decision-making process? Are area-based committees used, and if they are, is there a process of referral to a parent or council-wide committee? Can planning decisions be referred to full council?
- Are there any specific procedures within the local authority's constitution which impact on planning decision-making, such as area-based committees or reference of decisions up to a parent committee?
- Does the internal culture of the decision-making suggest a collaborative approach between members or does it tend to be more politically divided?
- What can be learnt from examining controversial planning applications? Are there particular types of application which generate controversy? What kinds of tactics do objectors typically employ? Are there any civic groups which regularly intervene in the process? Are these groups well-connected or influential in political circles?

Tactical considerations

- What opportunities exist for engaging with political representatives and at which stage of the process?
- How can elected members be involved in interactions with residents or other stakeholder groups?
- Should there be engagement with the leadership of the local authority?
- What materials and information should be supplied to elected members and what stage?
- How should objections be monitored and what is the process for responding to representations?

- Is there a time in the decision-making cycle when planning applications are likely to be determined in a more neutral environment?
- Are there any potential revisions to proposals that will address key political concerns?
- How can supporters be best involved in the political process and given the confidence to participate?
- Could there be a role for wider community surveys or independent polling to demonstrate the likely opinion of the wider community?
- Are there political representatives not directly involved in decision-making or influential civic groups who might be prepared to support a proposal?
- What are the key political messages and planning arguments that need to be relayed at the point of planning decision-making? Who should be involved in presentation opportunities either in a pre-application context or at the point of determination?

Conclusion

This chapter has examined the pre-eminent role that local government plays in the administration of the planning system. It has also considered the interface between officers and elected members and the political context in which planning operates. An easy mistake is to believe that the planning decision-making process is a technical one and planning merits are considered purely in terms of material considerations. Those involved in the planning system ignore the political and community environment at their peril and should develop a strategy to respond to it through effective community and political involvement.

Notes

1 The Stationery Office Limited (1997) Local Government in England, Scotland and Wales, Third Report of the Committee on Standards in Public Life, Paragraphs 288 and 289.
2 Planning Advisory Service and the Local Government Association (2013) *Probity in Planning for Councillors and Officers*. London: Local Government Association.
3 DCLG (Department for Communities and Local Government) (2014) *The National Planning Policy Framework*, Paragraph 187
4 Material and non-material planning considerations are listed in Appendix 2
5 RTPI (2016) *Code of Professional Conduct* [Online]. Available www.rtpi.org.uk/media/1736907/rtpi_code_of_professional_conduct_-_feb_2016.pdf [Accessed 12 October 2016]
6 The Stationery Office Limited (1989) *The Local Government and Housing Act*. London: HMSO.

7 Plymouth City Council (2013) *Probity in Planning, Planning Committee Code of Practice* [Online]. Available democracy.plymouth.gov.uk/documents/s49229/App%20A %20Planning%20Committee%20Code%20of%20Practice.pdf [Accessed 12 October 2016]

8 The Stationery Office Limited (2013) Strengthening Transparency Around Lobbying, Committee on Standards in Public Life. London: HMSO.

9 Cameron, D. (2010) *Rebuilding Trust in Politics* [Online]. Available www.theguard ian.com/politics/2010/may/11/david-cameron-speech-full-text [Accessed 12 October 2016]

12 Appeals and public inquiries

Throughout the United Kingdom, both applicants and third parties have a right to appeal against decisions made by local planning authorities on matters such as planning permission, prior approval and enforcement notices.

At a basic level, it might be argued that recourse to the planning appeals system belies community involvement, since community involvement aims to make development more acceptable to local communities affected by it. In this chapter we will examine this point further, consider the operation of the system itself and identify what opportunities remain for community involvement.

The appeal system in England and Wales

The Planning Inspectorate is the custodian of the planning appeal system and it is an executive agency of both the Department for Communities and Local Government and the Welsh government.

Following the refusal of a planning application, applicants are able to exercise their right of appeal, and an independent planning inspector appointed by the Planning Inspectorate will normally officiate. The exact procedure to be followed is usually determined by the inspector and can take the form of a paper exercise, where written representations are considered; an informal hearing, which is effectively a roundtable discussion; or a full public inquiry.

It has often been suggested by those opposed to planning applications that it is unfair that they are not able to appeal decisions made by local planning authorities. The argument is that it is a denial of natural justice that only applicants can exercise a right of appeal. Some UK politicians have supported the concept of a third-party right of appeal and this has been put forward by various non-governmental organisations, including Friends of the Earth and the Campaign for the Protection of Rural England. The Conservative Party's planning green paper, issued prior to the 2010 general election, posed the idea that 'The planning system should be reformed to allow for appeals against local planning decisions from local residents.'[1]

The then Shadow Planning Minister Bob Neil MP said at a seminar held to discuss the green paper,

> We have looked and learned from how third-party appeals work [in Ireland] and will work with the industry to introduce a mechanism to weed out malicious appeals, and to define exactly what we mean by third parties. . . . once you accept that there should be a right of appeal, then the logic asks why should it only be limited to one side? That is how our plans will provide symmetry and fairness to the system.[2]

The third-party right of appeal did not eventually feature in the Localism Act 2011 and has not been trailed as part of any proposed legislation at the current time. However, it could be considered as a natural extension of Localism in the planning system and may return to the political agenda in the future.

Call-in powers[3] (sometimes known as the Caborn Principles after the planning minister who put them in place) – when the secretary of state decides to take decision-making power on an application out of the hands of the local planning authority – also exist. This can happen at any time during the planning decision-making process, but in practice tends to be triggered when the local authority has decided that it is minded to grant planning permission. Time and space for the secretary of state to be able to decide whether to call in an application is conferred on certain types of development, such as that which takes place in the green belt, outside of designated town centres, World Heritage sites, on playing fields or on land which is subject to substantial flood risk.

The procedures in deciding an appeal in Wales are broadly similar to those in England, and the Planning Inspectorate is also an executive agency of the Welsh government. There are some powers of 'recovery' (see below) where a minister makes the ultimate decision.

Although power to call in is very general, government policy is to exercise it sparingly and only in specific circumstances. The grounds for call-in were originally set out in 1999[4] in response to a parliamentary question by the planning minister. The specific grounds relate to cases which, for example, may conflict with national policies, could have significant effects beyond the immediate locality, or give rise to substantial cross-boundary or national controversy. An additional potential reason for call-in was added in 2012[5]: identifying cases that may also impact on economic growth and meeting housing needs across an area wider than a single local authority.

It is often the case when a local authority has indicated that it is minded to grant a planning consent, that third parties make representations to the Planning Inspectorate or the secretary of state to exercise call-in powers. It is commonplace for commercial interests or local authorities affected by proposed development in a neighbouring authority to make such representations. Local objectors also regularly seek to attract the interests of the secretary of state in a planning application.

Similarly, it can be difficult for applicants to make an assessment as to whether such lobbying for call-in is taking place, although often this is apparent through a continuation of campaigning activities by objectors. All the same, techniques which have been employed to raise objections are usually brought into play again, such as using traditional or social media to encourage people or organisations to write to the Planning Inspectorate or secretary of state. The same applies where the parties are seeking the recovery of the decision by the secretary of state.

In cases where lobbying by objectors is likely, applicants or supporters of planning applications should consider defensive responses to prevent one-way lobbying traffic.

There is no legal obligation for the government to indicate any reasons why call-in powers have not been exercised in specific cases, and this has been established in the courts,[6] although the government announced in 2001[7] that it would adopt a new policy of giving reasons for a decision not to call in a planning application.

The exercise of powers to 'recover' an appeal by the secretary of state is where the inspector will still consider the key issues in the case but with the ultimate decision being taken at ministerial level. The ability to do this is enshrined in the Town and Country Planning Act 1990.[8]

Recovery of an appeal can take place at any point of the appeal process, even after an inquiry has taken place, but it cannot be done once an inspector has issued a decision. As with the grounds for call-in, specific circumstances are considered by the secretary of state in deciding whether to recover an appeal. These were set out in a written statement[9] in 2008 and largely relate to proposals with more than local significance; developments of a certain scale; significance for the delivery the government's climate change programme; or important or novel issues of development control; and/or legal difficulties.

Separate arrangements and powers apply in London and can be exercised by the London mayor. The case study below indicates how applicants can seek the intervention of the London mayor. Objectors may also seek mayoral intervention in major applications, although this is unlikely if it becomes obvious through early stages of the mayoral referral process that the Greater London Authority (GLA) is likely to be supportive.

In the latter instance, is not unknown for members of the London Assembly and campaigners to seek to influence the position of the Greater London Authority on individual planning applications. Essential Living's proposals at 100 Avenue Road in Camden were raised by London Assembly member Andrew Boff, who sought assurances from then London Mayor Boris Johnson that he would not take over the determination of the planning application at that site if the London Borough of Camden decided to refuse it. The Save Swiss Cottage campaign group also encouraged local residents to write directly to the mayor given their concern that the early stages of the GLA referral process had indicated support from the authority's officers.

Box 12.1 Case Study: intervention in a planning application by the mayor of London

Convoys Wharf, Lewisham

Convoys Wharf is a prominent Thameside site located in north Deptford within the administrative boundaries of the London Borough of Lewisham. Historically, it was a royal dockyard dating back to the 16th century, but was now largely a vacant site with the exception of a Grade II listed former dockyard building.

The 16.6-hectare site had been the subject of various development proposals over an extended period since 2002. The latest proposals submitted in April 2013 involved an outline master plan conceived by leading architects Farrells and promoted by Hutchinson Whampoa Properties (Europe) Ltd. A mixed-use regeneration scheme was proposed, including residential, employment, retail, community, cultural, leisure, hotel and restaurant cafe uses.

The London mayor has to be consulted on all planning applications in London that are of potential strategic importance.[10] These are generally known as referred applications and the criteria for referral includes development, which is:

- 150 residential units or more;
- Over 30 m in height (outside the City of London); or
- Proposed on green belt or metropolitan open land.

Once an application has been received by local authority and it meets the criteria established by the Mayor of London Order, the London borough is required to refer it. The mayor then has six weeks to comment on the application and make an assessment as to whether is in accordance with the policies of the London plan. This is known as Stage 1 of the process.

The application is then determined in the normal fashion by the local authority's planning committee, and following this, the application is referred to the mayor for final decision. This is known as Stage 2 of the process, and the mayor has 14 days to decide whether the local authority's decision should stand, to direct refusal, or to take over the planning application and become the local planning authority.

However, there is an additional provision within the 2008 Order[11] which allows for the applicant to approach the London mayor and request that the application be recovered, and this can happen before determination by the local authority. In the case of Convoys Wharf, Hutchinson Whampoa made such an approach some six months after the original planning submission to London Borough of Lewisham.

In response to the applicant's request, then London Mayor Boris Johnson decided to become the local planning authority, indicating to the council that although the borough had performed well in recent years in terms of housing and affordable housing delivery, in this case, the

> planning history over the long-term particularly over the past year or so indicates that the breakdown in the relationship between the council of the applicant is such that I am not satisfied that you will be unable to reach a timely planning determination in respect of the current planning application for this important site.[12]

The criteria that should be applied in justifying taking over the application is also contained in the order, which sets out three policy tests as follows:

- Development would have a significant impact on the implementation of the London Plan
- Development would have significant effects likely to affect more than one London borough
- There are sound reasons for intervention.

The first and final policy tests were deemed to have been met in the case of Convoys Wharf, and this resulted in a public hearing in March 2014. The procedure followed at the hearing is similar to that followed in any local authority, with the main difference being that the decision was in the sole hands of the mayor. Public speaking rights are available for both the applicant and objectors. In the case of Convoys Wharf, a number of objectors took the opportunity to address the hearing including the mayor of Lewisham, local civic groups such as the Pepys Community Forum, the National Trust and the local member of Parliament.

The eventual outcome was that Boris Johnson agreed with his officers in the GLA Planning Unit that planning permission should be granted.

The intervention by the London mayor in this way has been criticised by some as being politically driven. This has been reinforced by a view that with the Convoys Wharf proposal and others, such as the Mount Pleasant Post Office site and the Bishopsgate Goodsyard, the takeover of planning powers has been where the relevant local planning authority is under the control of an opposing political party. There is also a complaint that strategic interventions of this nature go against the principles of Localism. However, London's regional and strategic importance was recognised and retained when regional-based planning was abandoned in England after 2010 (see Chapter 8).

How this might now play out with the election of the Labour mayor, including his willingness to exercise Section 7 powers, all remains to be seen.

The appeal system in Scotland

In Scotland, there are two planning appeal and review mechanisms.

The first is a local review body which is available to review decisions made by officers under delegated authority.[13] The body comprises at least three elected members and offers the opportunity to reconsider decisions that have been made under a scheme of delegation. Applicants can only trigger the process and it can take the form of written representations, a hearing and a site visit if deemed necessary.

The second relates to decisions made by councillors. Here, appeals against planning decisions by applicants are made to Scottish ministers who appoint a reporter to consider and determine the appeal.

There are opportunities for people other than the appellant and the council to participate, although this will depend on the type of appeal and the procedures being followed. Representations made to a council on an application will be passed on to the reporter (by the council). Objectors to or supporters of a planning application will be given the chance to give further comments to the reporter and might also be invited to join in any further appeal procedures.

The reporter is appointed by Scottish ministers to make the decision on their behalf.

A very small number of appeals are not delegated to reporters for decision, but instead are 'recalled' by Scottish ministers who will then make the final decision. In those cases, and similar to the recovery process in England and Wales, the appeal will still be examined by a reporter, who will then write a report and make a recommendation for ministers to consider before they make their decision. As in England, ministers do not have to agree with the reporter's recommendation.

In general, an appeal will be recalled for ministerial decision only where it raises issues of genuine national interest. A substantial level of public interest in a development does not in itself provide sufficient grounds for an appeal to be recalled, although in its decision to recall Dart Energy's appeal in 2014 around proposals for the development of coal bed methane production within the administrative areas covered by Stirling and Falkirk Councils, the Scottish government cited 'considerable public interest in the proposals, as well as its relevance to the implementation of the new recently updated Scottish Planning Policy'.[14]

The appeal system in Northern Ireland

Northern Ireland has an independent appeals body, the Planning Appeals Commission (PAC). It has 89 functions under various pieces of legislation.

Applications for planning permission for certain major developments must be made to the Department for Infrastructure rather than to the local council.[15] The major developments concerned are those which have been identified by the department as being of significance to the whole or a substantial part of Northern Ireland, having significant effects outside Northern Ireland or involving a substantial departure from the local plan. The department can also give a direction requiring a planning application made to a council to be referred to it instead of being dealt with by the council. Such applications are known as called-in applications.

The commission is not part of any government department, although it receives financial and administrative support from the Department of Justice, its 'sponsor' department, which appoints the commissioners. Commissioners are not civil servants but are appointed following open public competition. The commission is wholly independent of its sponsor department in terms of decision-making and the operation of the appeals and hearing/inquiry/examination processes. At present, there are 17 commissioners.

Again, there is no third-party right of appeal against the planning decision, although when an application is subject to appeal, objectors or any stakeholders in the proposal may make a response to the PAC. The commission will write to all third parties notified to it as having made a representation to the local council and invite them to participate in the appeal process. It is then for these parties to decide whether to rely on their letter of objection or support and take no further part in the appeal process. They can, however, submit further written representations and/or attend any hearing or accompanied site visit that is arranged.

Box 12.2 Planning appeals in figures

The scale of activities around planning appeals can be discovered in statistics provided by relevant government departments.

In England, in the first quarter of 2016, there were 2,963 Section 78[16] planning appeals and some 119,700 planning applications during the same three-month period. This means that appeals represented about only 2.5% of planning applications made. Further drilling down into the numbers also shows that of the 2,963 appeals, only 90 resulted in full public inquiries. The rest were dealt with either by written representations or by informal hearings.

Statistics collected by relevant government departments reveal that:

- In 2015/16, there were some 11,812 S78 planning appeals received in England. This number is higher than in the previous five years, where the average was around 10,600.
- During the same year in England, around 89% of these appeals were to be dealt with by written representations, 7% by informal hearings and all 4% by formal public inquiries
- 10,214 appeals were decided in the same year in England. Of these, 9,227 were dealt with by written representations, 682 by hearings and 305 were the subject of a public inquiry
- Of the appeals decided during 2015/16, some 32% were allowed. 31% of appeals through written representations were allowed, 44% through hearings and 58% through public inquiries.
- The average percentage of appeals allowed over a five-year period in England was just over 33%
- In Scotland, there were 223 planning permission appeal cases decided in 2014/15, of which 118 were allowed (53%).

Judicial review

Although, as we have noted previously, there is no third-party right of appeal, there is the possibility of a challenge through courts following a planning decision made by a local authority or government department. The judicial review process exists primarily to give an opportunity for interested parties to challenge the procedure that has been followed in reaching a planning decision. It can focus on a decision, action or failure to act.

This is a tactic which has been regularly used by commercial interests or organised campaign groups in seeking to quash a public body's decision (known as an *order of certiorari*).

The process followed involves first notifying the defendant of an intended claim by letter. The purpose of the letter is to set out the key issues, which may be in dispute around the decision and to establish whether there might be ways of avoiding litigation. Claims for judicial review are dealt with in the planning court and challenges must be filed not later than six weeks after grounds to make the claim first arose. It is possible to challenge the secretary of state as well.

In the first instance, the claim for judicial review is considered by a single judge, based on the relevant papers. This stage is effectively an application for permission to proceed. If permission to proceed is refused or granted only on limited grounds, it is possible to seek a reconsideration of that decision at an oral hearing.

One of the problems with the judicial review process, which has been encountered by some applicants, is that they have not been aware of the claim being lodged, as it has not included them as a party.

Sometimes, the threat of judicial review or the granting of permission to proceed results in a local authority reconsidering planning applications to overcome procedural mistakes that may have been made.

Given that ultimately all costs are generally borne by the party losing the substantive claim at the end of the process, judicial review is often used as a tactic or a bargaining position. Individuals or residents' groups could leave themselves liable for substantial legal bills that have been run up by themselves and the defendants, which is why it is not a viable avenue for many.

In addition, even if a judicial review claim is eventually successful, it is possible for an applicant to return with an amended planning application that will navigate around the circumstances on which a claim was found to be valid. In this respect, it can be seen as more of a delaying tactic, although it is possible that a successful claim might have severe consequences upon the viability of proposals. A recent claim for judicial review is described in Box 12.3.

In major schemes, it is not unusual for applicants to share appropriate legal advice they may have received with local authorities to ensure that they do not make errors that could form the basis of a legal challenge. It underlines the importance of local authorities ensuring that all material planning considerations are carefully and properly considered and that internal and statutory administrative procedures are followed.

Box 12.3 Case study: judicial review

Grimsby Docks, Lincolnshire

SAVE Britain's Heritage is a group that has been has been campaigning for historic buildings since its formation in 1975 by a group of architectural historians, writers, journalists and planners.

In July 2016, SAVE launched a legal challenge to protect a series of historic port buildings at Grimsby Docks, Lincolnshire. The Victorian brick buildings, known collectively as the Cosalt Buildings on Fish Dock Road, form part of the historic Grimsby Dockyard and were the subject of prior approval for demolition by North East Lincolnshire Council. Associated British Ports were the applicants and the demolition of the buildings concerned were to facilitate the provision of new offshore wind facilities.

The basis of the judicial review was the council's screening opinion for an environmental impact assessment. SAVE claimed that this was unlawful because it did not correctly identify heritage assets on which there was likely to be an impact. The group maintained that the inherent value of the buildings as heritage assets was underlined by objections from Historic England (the government's advisers on heritage), the World Monuments Fund, The Victorian Society, Great Grimsby Ice Factory Trust and hundreds of people who signed a petition to protect the buildings.

As internal stripping out of the buildings in preparation of demolition had already started, SAVE also applied for an emergency injunction to halt further works whilst the judicial review process was underway. The group also launched an emergency appeal for financial support for the legal proceedings.

At the time, Henrietta Billings, Director of SAVE, said,

> These are not the first buildings in Grimsby Docks to be threatened with demolition and we are extremely concerned about the potential piecemeal destruction of such an important historic site. These buildings need urgent protection and rightful recognition and we are confident they can be retained and reused to regenerate this part of the docks, as part of a wider strategic approach to the site. We have been advised that we have good grounds for a legal review, and are keen to ensure that the demolition is robustly challenged.[17]

Appeals and the scope for further community and political involvement

In Chapter 11, we considered those issues which should be taken into account in terms of understanding the political decision-making context of individual local planning authorities. In terms of planning appeals, it is possible to identify individual local authorities that have enjoyed less success in defending their decisions at appeal.

Although it is important to be cautious with such statistics, it is possible to identify local authorities which may not be making sound planning decisions either on the basis of supported recommendations of officers or through a decision by elected members, where an officer's recommendation may be overturned. This may be helpful by way of background research, assisting those involved in development proposals in gaining an understanding of the planning decision-making environment which they might encounter in individual planning authorities.

Box 12.4 shows those authorities in England which are the least success-ful in defending their planning decisions at appeal, although the numbers of appeals which have taken place in each authority vary significantly.

It is also possible to look at the records of individual planning inspec-tors to understand how likely they are to allow appeals which they hear. Undoubtedly, members of the legal profession involved in planning appeals have their views on the propensity of individual inspectors to allow appeals and whether the chances of success are affected by which independent inspector is appointed.

Box 12.4 Comparative success rates in defending planning decisions at appeal

The statistics[18] shown in the table below highlight those English local authorities that have enjoyed the least success in defending their deci-sions through the planning appeal system.

Table 12.1 Comparative success rates in defending planning decisions at appeal

Local authority	No. of S78 appeals heard	No. allowed	% allowed
Arun	33	17	52
Blackburn with Darwen	13	7	54
Broxtowe	10	6	60
Christchurch	16	9	56
City of London	3	2	67
Dartford	18	9	50
Great Yarmouth	14	7	50
Halton	2	1	50
Hyndburn	4	3	75
Kettering	12	6	50
Mansfield	5	4	80
North West Leicestershire	14	8	57
Peterborough	17	11	65
Portsmouth	16	9	56
Richmondshire	6	3	50
Salford	5	5	100
Sandwell	9	6	67
Swale	37	22	59
Tendring	60	32	53
Thanet	17	9	53
The Broads Authority	4	4	100
Welwyn Hatfield	13	7	54
West Devon	27	14	52
Wigan	10	5	50

If the primary objective of community involvement in planning is to build consensus around planning policy or individual planning applications, it could be said that recourse to appeals represents a failure of that primary objective. Objectors often feel particularly aggrieved that once having persuaded their local authority that an application should be refused, they then have to continue the campaign through the appeal process.

Depending on the nature of the process that is followed with an appeal, objectors regularly put forward the argument, particularly in the case of public inquiries, that applicants can afford to employ the best professional advice, including leading counsel to advocate on their behalf. The claim is often made that 'the playing field is not level', as community-based objectors do not have the financial resources of commercial organisations. Clearly, if local authorities are defending a decision on a major development proposal at appeal then they will usually employ a legal team, and this might also be the case where commercial interests are seeking to oppose a planning application through an inquiry process.

Before an applicant decides to pursue an appeal, there is always a possibility of considering a resubmission of an amended planning application to address particular reasons why an application has been refused. The ability to do this depends or whether the reasons given for refusal are fundamental and/or cannot be addressed without losing the overall financial viability of a particular proposal.

One of the problems often encountered by applicants in these situations in seeking further dialogue with the local community or local politicians is that in the case of refusal, the objectors often have a mindset which is based on having 'won'. Is also often difficult to convince local residents that there is a case for further negotiation even if an appeal might be running. An argument which suggests that an appeal may result in a favourable outcome for the applicant and that a revised application may provide a more acceptable scheme often falls on deaf ears.

However, such further community political involvement on a revised scheme may be viewed as beneficial by all parties and it is an option that can be considered before going to appeal. It is also not unusual for neighbouring landowners to approach applicants who are considering an appeal to suggest ways in which a proposal might be modified. One of the principal considerations here, however, is whether there is any good prospect of the political decision makers backing a revised scheme. In cases where elected members have gone against the recommendation of their professional officers, applicants might simply conclude that their prospects are better at appeal and that the timescales involved in a resubmission may not be significantly quicker or more certain than pursuing an appeal.

Once the appeal process is underway, opportunities exist for interested parties such as local residents and civic groups to make representations.

The local authority itself will normally notify interested parties of the appeal start date and the timetable for the submission of any representations.

For objectors and supporters, there are tactical considerations such as whether they wish to take a more substantive role in an inquiry, perhaps by speaking at an inquiry or even submitting a statement of case. In some high-profile cases, objectors employ their own professional and legal advisers, funded by local community fundraising campaigns.

Supportive applicants will need to consider whether and on what basis they might involve individuals or groups. In practice, the most usual intervention within the appeal process by members of the local community is through submission of written representations supplementing those that may be made around the consideration of the application by the local authority. Campaign groups objecting to development often contact their supporters to encourage them to write to the Planning Inspectorate, on the basis that a large volume of objections will help influence eventual outcome.

The concept of a planning application ending up at appeal due to a failure of community involvement is an oversimplification perhaps, since it is possible to build community consensus around planning interventions, yet still end up with a local authority choosing a different site allocation within a local plan or refusing a planning application.

It can also be the case – as we have seen in Chapter 11 – that local opinion or pressure on political decision makers can have a significant impact in reaching a balanced judgement of the planning merits. It is probably correct to say that in the vast majority of cases, a lack of community support or consensus is usually critical in determining the stance of political decision makers, particularly where a proposal may have vulnerable or marginal aspects when considered against material planning considerations. Conversely, strong community support can swing the balance the other way.

There is also a further possibility in that central government can decide to intervene and 'call in' a planning application, which may in fact enjoy both community support and also backing of the local planning authority.

An example of such a situation was with a proposed new designer outlet centre at Scotch Corner in Richmondshire, North Yorkshire. Local community involvement activities revealed significant local support from both the public and community-based organisations such as parish councils. Richmondshire District Council resolved to grant planning consent, following a positive recommendation to do so by the local authority's officers. There was, in the council's view, both local support and also compliance with relevant planning policies relating to retail development. However, the secretary of state decided to call in the proposal and this led to a public inquiry.

In this instance, interested parties such as residents, civic groups, pressure groups, councillors and MPs may make representations directly to the secretary of state and the Department for Communities and Local Government ministers involved in planning decisions.

The DCLG publishes guidance[19] regarding how ministers need to conduct themselves in relation to representations on planning matters. In particular, planning ministers are required to demonstrate evenhandedness and provide to all parties any evidence which is material to any decision that has been the subject of a planning inquiry. The secretary of state is also required to allow parties a further opportunity to make representations if, at the close of an inquiry, the secretary of state differs from the inspector regarding any relevant matter of fact or proposes to take any into account any new evidence. The potential for a judicial review is taken into account here, highlighting the risk that allowing privileged opportunity to any party to make representations may be deemed procedurally unfair.

There is specific reference in the guidance to call-ins. As indicated previously, it is often the case that planning ministers are asked to call in planning applications where the local authority is likely to be supportive. According to the guidance, those seeking to make such representations should be directed to relevant officials of the department. It also states that where representations are made by whatever means, including letters, telephone or email, whether directly to a planning minister or to an official, it should be indicated that such representations can only be taken into account if they are also made available to all interested parties for comment.

Overall, although the operation of the appeals system generally means that scope for further community involvement is limited due to its adversarial nature, this chapter has indicated that stakeholders have opportunities to participate in the process. For applicants, as we have considered, there may be scope to involve supporters in the process or make representations at a strategic political level.

Notes

1 Conservative Green Paper No.14, *Open Source Planning*.
2 As quoted in the March 2010 House Builders Association briefing.
3 The Stationery Office Limited (1990) *Town and Country Planning Act* Section 77
4 Hansard (1999) House of Commons Debates 16 June 1999 c138w
5 Hansard (2012) House of Commons Debates 26 October 2012
6 R v Secretary of State for the Environment, ex p. Newprop [1983]
7 Hansard (2001) House of Lords Debates 12 December 2001
8 The Stationery Office Limited (2008) *Planning Act* Section 79
9 Hansard (2008) House of Commons Debates 30 June 2008 c44WS
10 The Stationery Office Limited (2008) The Town and Country Planning (Mayor of London) Order
11 Ibid., Section 7
12 Johnson, B. (2013) in a letter to the Lewisham Council Planning Department 30 October 2013
13 The Stationery Office Limited (2008) The Town and Country Planning (Schemes of Delegation and Local Review Procedure) (Scotland) Regulations

14 Scottish Government (2014) *Planning Appeals Recalled 10 October 2014* [Online] Available news.scotland.gov.uk/News/Planning-appeals-recalled-113e.aspx [Accessed 12 October 2016]
15 The Stationery Office Limited (2011) *Planning Act (Northern Ireland)*, Section 26
16 The Stationery Office Limited (1990) *Town and Country Planning Act*, Section 78 as amended
17 SAVE Britain's Heritage (2016) Press release: SAVE challenges demolition of historic Grimsby Docks buildings and launches fundraising appeal press release, 8 July 2016 [Online] Available www.savebritainsheritage.org/campaigns/item/394/Press-release-SAVE-challenges-demolition-of-historic-Grimsby-Docks-buildings-and-launches-fundraising-appeal [Accessed 12 October 2016]
18 Planning Inspectorate (2015) PINS Business Intelligence System, June 2016, provisional figures for April 2015 to March 2016
19 DCLG (Department for Communities and Local Government) (2012) *Guidance on Planning Propriety Issues*. London: DCLG.

13 Consulting on a nationally significant infrastructure project

Introduction

The process of consulting on nationally significant infrastructure projects (NSIPs) differs considerably from both Localism's bottom-up approach and the standard planning application process.

The definition of an NSIP is set out in the 2008 Planning Act: essentially an NSIP is an infrastructure development above a specific threshold which is identified as being of national importance. Typically, this will include railways, highways, water, waste, harbours, power-generating stations, wind farms and electricity transmission lines.

Such projects typically span several local authority areas, are complex, are frequently controversial and often involve compulsory purchase. In response to these limitations – and importantly the fact that these projects are of national significance – the planning system allows for consultation on the principle of the development to occur at a national level, while the local consultation is centred on design, community benefits and mitigating negative impacts to the neighbourhood. This alternative – often described as 'top-down' – approach is intended to be faster and smoother and introduce more certainty to the planning process.

The specific procedures for NSIPs were set out in the Planning Act, which also introduced an Infrastructure Planning Commission. Two years later, the National Infrastructure Plan led to the abolition of the Infrastructure Planning Commission. Responsibilities were transferred to the Planning Inspectorate and the approvals process transferred to the secretary of state through a development consent order (DCO).

A National Infrastructure Commission (NIC) came into being in 2015. Its first priority was to focus on three priority areas: the connectivity of the northern cities (including a potential HS3), large-scale investment in London's public transport infrastructure and increasing efficiencies in energy infrastructure investment. Other priorities will come into play during the lifetime of the NIC.

Table 13.1 The NSIP process

	Developer	Local authority	Planning Inspectorate	Secretary of State	Time limit
Step 1 Preparing a Statement of Community Consultation (SoCC)	Informs the Planning Inspectorate of a wish to submit a planning application Draws up SoCC, setting out the consultation strategy and timetable	Inputs informally into the SoCC			
Step 2		Approves the SoCC			28 days
Step 3 Pre-application	Carries out a comprehensive consultation in line with the SoCC	Oversees the consultation, assisting as appropriate and ensuring that the SoCC is followed closely			
Step 4 Acceptance	Submits an application to the Planning Inspectorate		Decides whether the application meets the required standards		28 days
Step 5 Pre-examination			Invites the public to register with the Planning Inspectorate and submit written responses Holds a preliminary meeting, which may be attended by all those registered		3 months

Step 6	Examination		Leads an examination and in some cases a public hearing	6 months
Step 7	Decision		Makes a recommendation to the secretary of state	3 months
Step 8			Issues a decision on the proposal	3 months
Step 9	Post-decision [legal challenges may occur at this stage]			6 weeks

The NSIP planning process

Permission of the principle, but not the siting, of an NSIP exists in the form of a national policy statement (NPS). There are six energy NPSs,[1] three transport NPSs[2] and three water and waste NPSs,[3] all of which have been instigated by the relevant secretary of state, consulted upon at a national level and approved by Parliament. In the case of nuclear power, however, the general location of a specific development is determined through the NPS. Where this is the case, the secretary of state is required to take 'appropriate steps' to publicise the draft NPS locally and must consult all relevant local authorities.

With the principle for the development already established, the process of submitting a development consent order application under the NSIP process is therefore a fairly straightforward – albeit substantial – one.

Following a positive decision and in the absence of any legal challenges, consent in the form of a development consent order, together with rights to compulsorily purchase land, is granted.

Although legitimacy of the development is secured prior to the applicant's involvement, and despite the process being defined and streamlined, NSIP applications are not without their challenges – not least the extremely comprehensive and prescriptive level of consultation that is required.

Stakeholders

Any NSIP consultation will involve stakeholders from a variety of backgrounds, interests and locations. In some cases, such as with rail or road applications, stakeholders will cover a wide geographic area.

Under Section 42 (Duty to Consult) of the Infrastructure Planning (Applications, Prescribed Forms and Procedures) Regulations 2009, all local authorities in the vicinity and those with an interest in the land must be consulted. Section 47 (Duty to Consult the Local Community) requests that the developer, together with the local authority, produces a Statement of Community Consultation which clearly states the way in which the broader community will be consulted.

Research, stakeholder mapping, and sustained contact is of particular importance due to the diverse nature of stakeholders involved and the emphasis on thorough consultation and reporting.

The SoCC

The Statement of Community Consultation is a comprehensive document produced by the developer which sets out the proposed consultation strategy and tactics. The local authority is obliged both to comment on the proposed approach, as the SoCC is developed, and also to approve the SoCC formally at the end of the process. A local authority usually provides invaluable insight into the specific groups which make up the community and the most appropriate means of reaching hard-to-reach groups, and must also alert the developer to any comments it has received from local residents concerning the process of consultation.

Typical contents of a SoCC are as follows:

1 Introduction
2 Project context
3 Project description
4 National planning context
5 Consultation process for NSIPs
6 Local planning context
7 Section 42 consultees
8 Section 47 consultees
9 Local bodies and representative groups to be consulted
10 Publicity
11 Reporting and evaluation
12 Consultation materials
13 Issues to be addressed
14 Suggested location and timings of public events
15 Timetable
16 Appendices – such maps of consultation zones, electoral wards and proposed routes, along with any feasibility studies carried out in relation to the project

Once the SoCC has been advertised, very few changes may be made by the developer without drafting, gaining approval for and publicising a new SoCC. Generally speaking, it is possible to increase the extent of the agreed consultation, but not to reduce it.

The consultation

An approved SoCC will provide the necessary template for a comprehensive consultation. As with any consultation, the principles of best practice[4] should guide the strategy: it should be strategic, balanced, responsive, genuine, engaging and timely and seek to learn from the local community while also managing residents' expectations.

NSIP consultations may differ from other consultations in their very rigid approach, which can provide less opportunity to adapt to changing stakeholder demands and less opportunity for creativity. However, an investment in research and dialogue in the early stages of the process should ensure that the consultation strategy is well suited to the project and requires little change.

One of the most difficult issues for an NSIP consultation is at its very heart: the fact that the consultation does not seek to address the need for the development. The consultation addresses not *whether* the development goes ahead but *how*. Invariably, the local community will wish to vocalise views on the very principle of development, but this is outside the remit of the consultation. This inevitable issue should be addressed at an early stage through an explanation of the process, stating the aims of the consultation clearly and managing expectations. Although the question of need is not within the remits of the consultation, a good consultation will communicate purpose of the development to ensure that the community is fully acquainted with the reasons for the change. Engaging with the public at such an early stage is beneficial in building trust. Therefore, an applicant of an NSIP should seek to compensate for the absence of this early-stage engagement by running as comprehensive, transparent and fair a consultation as the process allows.

The question of need provides an opportunity to communicate the purpose of the planning application. For example, an applicant seeking to gain a DCO for a power station faced with the comment, 'I don't want those chimneys obscuring my view', would be advised to engage on the *need* for the scheme at an early stage – in this case, addressing the need for this particular source of power and the benefits that it brings to the community.

It should also be accepted, however, that many infrastructure projects have immediate negative features, and a good engagement process will ensure that these are outweighed by benefits, both direct and indirect. Understandably, many people are unaware of the need for infrastructure, particularly in specific circumstances, although in reality most national infrastructure is of relevance and benefit to a wide range of people. It takes a well-thought-through and consistent communications strategy to communicate this effectively.

While the principle of the development is not a subject for debate, many subjects are open for discussion. Broadly, these include:

- Design and environmental sustainability
- Impact on the neighbourhood, its existing infrastructure and facilities
- Community benefits to be provided both within the scheme and in the wider neighbourhood
- Construction practices and measures to mitigate the adverse impacts of the development such as landscaping, noise attenuation and hours of operation.

With any large project, there is a danger that the project can be politicised. Particularly in the run-up to a general election, MPs and councillors may campaign strongly in favour of or against the scheme, causing emotions to run high.

Another potential difference is a very rigorous approach to feedback. Collation of results, analysis, reporting and presenting the local feedback must be comprehensive, detailed and scientific. The applicant's response to each of the issues raised must be shown clearly and must demonstrate whether the development proposal has been adapted in light of the comments.

The consultation report

The application for an NSIP must be accompanied by a consultation report in accordance with Section 27 of the 2008 Planning Act, which states that the report must include:

1 An account of the statutory consultation, publicity, deadlines set, and community consultation activities undertaken by the applicant at the pre-application stage under s42, s47 and s48;
2 A summary of the relevant responses to the separate strands of consultation; and
3 The account taken of responses in developing the application from proposed to final form, as required by s49 (2).

The report should be focused and concise, but comprehensive too. Typical contents are as follows:

1 An overview and narrative of the process, including a summary of consultation activity. This section should cover the planning and consultation history of the project.
2 The policy context, quoting directly from the relevant NPS.
3 A project and company overview.
4 A definition of the vicinity of the consultation.
5 A list of those consulted including landowners, local authorities and other statutory consultees. The definition of statutory consultees is set out in Schedule 1 of the Applications: Prescribed Forms and Procedures Regulations 2009, and any variation from these recommendations should be explained. Ideally, the report will include an explanation as to the stage of the consultation at which each stakeholder group was involved and the specific tactics used to reach them.
6 An overview and summary of the consultation activities. This should mirror the content of the SoCC. Any inconsistencies such as activities not included in the original SoCC should be explained.
7 Information relating to statutory publicity notice as it appeared in the local and national newspapers together with the timing of the publicity. The report might also include a description of the consultation material used and how the prescribed consultees were able to access it, although this is not compulsory.
8 Details of any non-statutory 'informal' consultation.
9 The Environmental Impact Assessment (EIA) Regulations consultation where this forms part of the overall consultation.
10 The method of reporting and analysis.

11 Statutory consultation responses, summarised by issue and presented both qualitatively and quantitatively. Names and addresses of those consulted must be included, along with analysis on the basis of ethnicity, age and socio-demographic grouping.

12 A summary of responses, stating what changes have been made in relation to the issues raised. These should be grouped as follows: Section 42 prescribed consultees (including Section 3 and Section 44), Section 47 community consultees and Section 48 responses to statutory publicity.

As the consultation report will be published on the Planning Inspectorate's National Infrastructure portal, developers should ensure that it complies with the Data Protection Act 1998. Personal information such as names and addresses must be used only for the purpose of the consultation report and information must be redacted as appropriate.

The role of the local authority

The formal role of the local authority has been addressed. However, local authorities should also be mindful of their roles both in helping the developer to understand and navigate the local community and to help local people respond to the consultation where necessary. In both cases, the local authority can benefit the consultation, while also helping to manage expectations. Local authority planning and consultation teams are advised to develop communications strategies in relation to the proposed development. Strategies should seek to open up channels of communication, particularly for those less likely to become involved of their own accord, enabling better dialogue both with the developer's consultation and also with elected representatives who have a responsibility to understand and represent their constituents' concerns. A good strategy will take into account broader policies within the local authority and ensure that the council's position on the proposed development does not conflict with other priorities.

In many large-scale projects, more than one local authority will be affected, and therefore it is necessary for local authorities to communicate among themselves and with other democratic bodies, such as neighbourhood forums and parish councils.

While ensuring that the consultation is comprehensive and wide reaching, the local authority should also be aware of the dangers of consultation fatigue, particularly for projects with a long duration. The local authority will be best placed to advise the developer on how specific groups within the community can be targeted most effectively and efficiently.

Community benefits

Residents, developers, public bodies and indeed the NSIP process itself acknowledge that no development is possible without disruption to a local community and that particularly in the case of some very contentious NSIPs, drawbacks (perceived or otherwise) should be balanced with benefits. Thus, every NSIP invests substantially in its location. Typically, community benefits are either physical infrastructure (such as educational facilities, visitor recreation facilities and the protection of natural habitats), funding (for local events, community facilities and environmental initiatives) or financial (shares in profits).

This presents a real opportunity for genuine participation, with the developer working in partnership with local residents to develop a package of benefits which respond to local needs and aspirations, celebrate the local area and create a lasting legacy. Developers will often find it helpful to begin the process with a deliberate event, such as a charrette or weekend workshop; to set up a community benefits liaison group, consisting of residents, businesses, local politicians and representatives from the LEP etc.; to put in place a regular form of communication with the wider public, such as a newsletter or dedicated website; and to hold regular update meetings.

In the case of wind farms, developers must complete the English Register of Community Benefits – a database which enables developers to record the benefits they have voluntarily offered to communities.

Box 13.1 Case study: NSIP consultation

Arup – Thames Tideway Tunnel

At present, untreated sewage mixed with rainwater regularly overflows from London's Victorian sewerage system into the River Thames via combined sewer overflows, failing to conform to relevant European wastewater legislation. The solution to the overflows has been examined for a number of years, and the Thames Tideway Tunnel was put forward as the solution. The project was granted a development consent order in 2014.

Thames Water appointed Arup to work with them on the statutory consultation associated with preparation of an application for consent for the scheme. This commenced prior to the project being designated a NSIP (in 2012), but was undertaken in anticipation that a designation would be forthcoming, as indicated by Defra in November 2009.

The SoCC

Central to the pre-application process was preparation of a consultation strategy. There were multiple rounds of consultation on the draft stakeholder and community engagement strategy (at this point the 2008 Act had not come into force), then the Statement of Community

Consultation (as required by the 2008 Act) and community consultation strategy (a more detailed informal engagement strategy), from autumn 2008 to summer 2011. Each round sought feedback from those local authorities directly affected by the project, neighbouring authorities and statutory consultees likely to have a strategic interest in the project. The aim was to ensure that the approach to statutory consultation was informed by local and strategic interests and the findings from each stage of engagement were reported to demonstrate how the document had evolved in response to feedback. Given the scale of the project, each stage of engagement exceeded the statutory minimum of 28 days.

In accordance with Section 47(6) of the 2008 Act, the SoCC was published in prior to the two stages of consultation. Subsequently, the consultation report demonstrated how the statutory consultation and Section 48 publicity was completed in accordance with the SoCC.

Pre-application consultation

The pre-application consultation began in September 2010 in five clearly defined steps:

1   Phase one consultation (September 2010 to January 2011 – 18 weeks). This stage of the consultation involved statutory consultees, local authorities, landowners and community consultees and received feedback from 2,869 respondents. The content of the consultation addressed:

- The need for the project
- The alternatives to a tunnel solution
- Findings on the preferred scheme, consisting of a series of preferred sites and the preferred route, alongside other shortlisted sites and routes
- Engineering, environmental, community, planning and property issues that had been identified
- Initial ideas for permanent structures and the use of each preferred site after construction work.

At the end of this stage, a report on Phase One consultation was published.

2   Interim engagement (March to October 2011 – 22 weeks). Interim engagement was focused on community consultees and received feedback from 315 respondents. The issues covered were two sites where the intended outcome had changed from that previously

consulted upon, and nine sites identified as potent alternatives to those consulted on previously.

3   Phase two consultation (November 2011 to February 2012 – 14 weeks). This stage involved statutory consultees, local authorities, landowners and community consultees and received feedback from 6,019 respondents. It considered:

- The need for the project (to ensure that any new consultees had the opportunity to comment on this aspect of the project)
- The preferred tunnel route
- Preferred sites for construction and permanent works
- Detailed proposals for preferred sites
- The effects of the project as reported in the preliminary environmental information report.

4   Post Phase 2 consultation was undertaken in four parts between June and November 2012 and received 142 responses. This targeted consultation was focused on amendments proposed following the Phase 2 consultation. The SoCC had made allowance for targeted consultation and it was not necessary to further consult on the document at this stage.

5   Finally, publicity undertaken in accordance with Section 48 of the Planning Act 2008 (July to October 2012 – 12 weeks), providing:

- A summary of the main proposals, including the location and route of the proposed development
- Documents, plans and maps, which showed the nature and location of the proposed development.

83 respondents provided feedback on the Section 48 publicity.

Throughout the consultation, a variety of mechanisms were adopted to engage with stakeholders. The approach was fully compliant with the requirements of the 2008 Act and also aimed to make engaging with the project accessible.

At each stage of consultation, bespoke information was prepared, with a focus on ensuring that materials were accessible in terms of language and illustrations as well as easily available in range of formats and at a variety of locations, from public libraries to online. This included site information papers for each of the 26 sites and 17 project information papers, providing accessible information on project-wide topics. These were available, free of charge, on CD or in hard copy.

A programme of public exhibitions took place, over 49 days and 23 venues during Phase 1 of the consultation and over 57 days and

23 venues during Phase 2. During the targeted consultation, exhibitions were held over eight days in four venues. The number of venues reflects the linear nature of the project and the need to ensure that the exhibitions were held locally to communities where works were proposed, as well as in central London locations for those with a general interest in the project. All exhibitions were staffed by the project so visitors could ask questions and seek any clarifications.

A consultation website ran throughout, providing information on the project and each stage of the pre-application process. It included all materials produced to support each stage of consultation and linked feedback forms directly to the publicity materials to aid users. The website was designed to meet best practice standards in terms of accessibility and usability and encouraged feedback online. The feedback submitted online linked directly to the consultation database used to analyse all responses.

Accessibility

Throughout the consultation, it was important to ensure access for all. Consultation information was available in large print, Braille or audio format upon request and a customer centre offered a telephone service to translate consultation materials into any language on request. Health and safety audits of all exhibition locations were carried out and those selected were deemed accessible.

Feedback

Consultees were invited to submit written feedback on the proposals via feedback forms, correspondence (emails or letters), comment cards and the website. All feedback received equal weight. Arup applied a comprehensive analysis methodology to ensure that all feedback from each individual respondent was recorded, analysed and reported; this included a detailed system of coding for qualitative responses.

All feedback was analysed thematically, either for project-wide issues or for each site, to ensure that similar comments were grouped and considered together, consistent with guidance from the Planning Inspectorate. This approach allowed readers to have clarity on the site-specific comments, which was particularly important for this project, given it extended to 25 sites during consultation.

Processing and analysing feedback

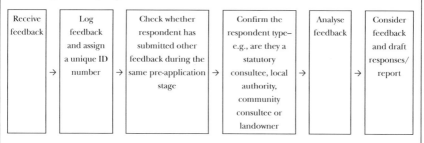

Receive feedback	Log feedback and assign a unique ID number	Check whether respondent has submitted other feedback during the same pre-application stage	Confirm the respondent type– e.g., are they a statutory consultee, local authority, community consultee or landowner	Analyse feedback	Consider feedback and draft responses/ report

After each stage of consultation and interim engagement, Arup:

- Held workshops with the project team to review the feedback received, identify the key issues and establish how the project would be amended (if appropriate) and the response to each issue.
- Prepared a report which set out the feedback received and the initial response to the feedback received, which was made publically available on the Thames Tideway Tunnel website and in hard and electronic copy on request (free of charge).
- Reviewed the effectiveness of the engagement to identify lessons learnt that could be applied to the next stage.

The DCO application was accepted in March 2013. Central to the acceptance of the DCO was the consultation report, which documented all stages of pre-application consultation and other engagement and is a statutory requirement for this type of application. The project was granted consent in September 2014.

Spanning 25 sites and 14 local authorities and taking into account almost 10,000 responses, this was a substantial consultation. The fact that it took place before, during and after the NSIP designation was put in place, requiring the project to make changes to the consultation as legislation changed, further complicated the consultation. The project's success is a result of collaborative working across diverse teams, flexibility among teams as required (which included extending teams to take on substantial workloads), comprehensive systems to ensure consistency and a well-managed system of processing feedback.

Conclusion

Although there are limitations on NSIP consultations in terms of remit and flexibility, significant opportunities exist for constructive local dialogue. Attention to specific community benefits along with a real understanding of, and a genuine desire to mitigate, the downsides of development, can significantly benefit a scheme – not just at the planning stages but over its lifetime.

Notes

1 Overarching energy, renewable energy, fossil fuels, oil and gas supply and storage, electricity networks and nuclear power.
2 Ports, airports and national networks.
3 Water supply, hazardous waste and wastewater treatment.
4 The principles of consultation are described in detail in Chapter 14.

Part III

Communications strategy and tactics

14 Strategy development

Introduction

Strategy is the basis upon which a communications programme is formulated: the means by which the general direction, the selection of tactics and the timing of a consultation are decided.

A clear strategy is the starting point for a successful consultation. The logical sequence of a strategy, however wide-ranging the involvement activities, establishes clear aims and objectives, enabling the consultation team to share values, expectations and understanding. It is also the best means of identifying relevant factors at the start of the process and taking these into consideration as the consultation progresses.

This chapter outlines a comprehensive approach to strategy development. Whether this is followed or not, it benefits the consulting body to have awareness of the comprehensive approach. A reduced version can be found in Appendix 3.

The strategic approach

Formulating a strategy, as this chapter will show, need not require substantial understanding of communications theory or days spent grafting and crafting. And the benefits are numerous. In many planning applications – such as those on NSIPs or carried out by a public body – a clearly defined strategy is required by law. This alone speaks volumes as to its importance. Although a strategy may not be a legal requirement for other planning applications, NSIP applications require a consultation report[1] which details the approach taken, an evaluation of the results, and decisions on the future direction of the project, based on consultation responses. The success of each of these is dependent upon there being a coherent strategy.

Transparency – always an important characteristic of consultation – is another benefit of a strategic approach. Without it, the launch of a consultation may be met with confusion. It is recommended that, at the implementation of a consultation, the strategy is communicated in a consultation mandate which sets out the vision, objectives and structure of a consultation and shows how results will be received, analysed and acted upon. This can avoid confusion among third parties, whether special interest groups, local residents, statutory consultees or public bodies.

A strategic approach also allows for resources – whether financial or human – to be allocated and planned, thus enabling the consultation to be run efficiently.

A common mistake in planning, often despite better intentions, is for a strategy to become a retrospective document: the team launches into a series of consultation tactics (perhaps based on past practice, experience or recommendation), results are collated, and then in a need to create a meaningful consultation report, a 'strategy' is drafted to justify the approach.

Worse still, and all too common to the industry, is to 'predict and provide' (to make assumptions about what a development should comprise and put the proposals in place with little or no consultation); 'plan, announce and defend' (to put in place a development proposal, inform local residents and attempt to counter any negative sentiment); or to 'plan, monitor, manage' (to put in place a development proposal, gather opinion and then attempt to promote only the positive opinion). Each of these examples is a distinctly top-down approach and makes scant use of consultation. Of course, developments require varying levels of public consultation, but those which are seen to be avoiding any meaningful dialogue are setting themselves up for failure.

If it be said that strategy is a waste of time, it is only in the case of a retrospective or tokenistic 'strategy' that this is true. And in reality a good strategy is never retrospective or tokenistic.

Differing approaches to consultation strategy

Planning consultations vary hugely. The numbers of those involved can range from an application for a handful of residential units in an underpopulated area to a 10-year local plan for a city involving thousands of stakeholders including both existing and future residents. Similarly, the length of the consultation might be anything from as little as six weeks for a one-off application for a community facility to the continuous process of strategic planning.

It should not be assumed, however, that a greater degree of consultation is of greater benefit: quite often the reverse is true. The case study below outlines two very distinct approaches to consultation which were undertaken by the same company. Both were regarded as successful and achieved planning consent.

Box 14.1 Case study: differing approaches to consultation strategy

Essential Land/Essential Living

In 2012, the London Borough of Bromley granted planning consent for the £150 million redevelopment of the former GlaxoSmithKline site in Beckenham, South East London. This followed a substantial programme of community engagement sustained over 15 months, which involved a community planning day and feedback session, the creation of a stakeholder group which met regularly throughout the process, a dedicated website, Facebook and Twitter pages, and many meetings with local groups. The community planning event, which included charrettes and visualisations by local residents, was well attended and welcomed by the established and highly engaged community. The comprehensive community consultation led to a community-inspired vision which was instrumental in planning consent being granted unanimously.

The Perfume Factory, a mixed-use, predominately build-to-rent scheme which utilises an industrial/commercial building in the London Borough of Ealing, followed a very different approach to consultation. A largely industrial area with some student housing but very few homes, the location lacked an active local community. Furthermore, the designation of the Park Royal Opportunity Area in which the scheme is located had already led to significant levels of consultation with the wider community. Essential Living held a public exhibition and communicated via social media and a dedicated website, but response rates were relatively low suggesting a general acceptance of the proposals among local residents. The Perfume Factory received resolution to grant-planning consent in February 2016.

As the examples show, consultation strategies can differ significantly. A consultation which aspires to high levels of involvement necessitates an open brief, the flexibility to introduce diverse features and the opportunity to make a real difference to a neighbourhood. It requires the audience to be interested, motivated and open-minded. Adequate information must be provided and the development team must be willing to both adequately inform local residents and accept the outcome of the consultation, however unexpected. A more comprehensive approach is suitable in residential locations, particularly those with an established community, on land which has previously been used by the community, and in circumstances where a wide range of options is available for the site.

A more simple consultation is appropriate where the development proposal is unlikely to arouse significant interest, resources are limited, options are restricted due to viability issues, or the complex nature of the proposal requires an emphasis on professional input over local sentiment.

Invariably, the extent of consultation will be determined partially by the local authority, depending both on the local authority's interest in the scheme and its attitude towards engagement. Neither of the examples above is right or wrong: most importantly, consultation must be appropriate to the development proposal, meet the requirements of the public bodies involved, engage those who wish to have a voice and take local expectations into account. While a consultation which is too narrow can cause consultees to feel 'boxed in' and unable to express their thoughts, a more open approach can sometimes make it harder for residents to respond meaningfully.

The strategic framework

The formation of a strategy, therefore, is unarguably the first stage of the consultation process.

Table 14.1 shows how the important elements of consultation fit neatly into the strategic structure recognised by communications professionals as the most effective means of conducting a communications campaign.

Figure 14.1 shows how each stage of the strategy informs the next and the importance of continual strategy development:

Table 14.1 Planning requirements and strategy development

Function	Planning requirement		Stage of strategy development
Research	Local intelligence to benefit research	→	Pre-consultation dialogue
	Understanding the locality and the issues impacting on the suggested changes	→	Situational analysis
	Identifying local groups/individuals and compiling contact data	→	Stakeholder/publics analysis and research
Plan	A clear direction	→	Aims and objectives
	Consistency of voice	→	Messages
	The proposed approach	→	Strategy and resource allocation
Consult	Carry out the consultation	→	Tactics
Evaluation and analysis	Collate and interpret the results	→	Analysis
	Assess the effectiveness of the consultation	→	Evaluation
Feedback	Write consultation report and update community	→	Reporting and feedback

Figure 14.1 Stages of the strategic process

This chapter will address each of the fundamental stages of a consultation strategy: situational analysis, stakeholder/publics analysis and research, objectives, messages and strategy; Chapter 15 will provide guidance on selecting tactics and Chapter 17 will provide advice regarding analysis and tactics and evaluation.

Initial research

Pre-consultation dialogue

It is common for a consultation team to compile a consultation strategy internally, using research gathered informally from a variety of sources.

However, in many cases, specifically those which focus on ongoing involvement with a specific community – such as the process of consulting on a local plan, or the consultations carried out by housing associations which form part of a wider, continual, involvement exercise – a recommended starting point is to engage in dialogue with local residents prior to embarking on the consultation process. An NSIP application requires substantial pre-consultation engagement prior to producing a Statement of Community Consultation, and this approach has been seen to be very effective.

Situational analysis

This early stage of strategy formation should assess the context of the consultation as broadly as possible to ensure that all factors are taken into account.

Useful methods for ensuring thorough analysis are the PEST (political, economic, social, and technological) and SWOT (strengths, weaknesses, opportunities, threats) methods, which should identify both issues affecting the development proposal and also the consultation.

Table 14.2 Examples from a typical PEST analysis

	Development proposal	Consultation
Political	The proposal sits comfortably with the leading party's stance on housing provision for the borough.	Local elections within a year may mean that councillors will use the consultation as a platform for political campaigning.
Economic	CIL contributions are likely to raise a significant sum for local infrastructure, potentially enabling the citing of a new river crossing.	Resources are limited to 2% of the overall development budget.
Social	The development proposals include community facilities which would are much needed and as such as expected to be welcomed.	The local community has been consulted several times in the past year and consultation fatigue may affect levels of engagement.
Technological	The establishment of the new community is expected to bring about high speed broadband, not just to the new homes but to others in the village.	The local authority has a growing interest in focusing dialogue online, specifically in the case of planning.

Table 14.3 Examples from a typical SWOT analysis

	Development proposal	Consultation
Strengths	The proposal brings about much-needed housing, both private and affordable, and as such appeals to a wide cross-section of society.	The developer has a good reputation for comprehensive consultations, and its previous consultations have received positive comments from the local authority.
Weaknesses	The site has a history of failed planning applications.	The development proposal is intended to create affordable housing for specific groups which do not currently exist within the local community.
Opportunities	The population increase has the potential to reverse the decline of the high street.	The neighbourhood forum, which has excellent links into all sections of the community, can be used to reach a good cross-section of the community.

	Development proposal	*Consultation*
Threats	A planning application for a neighbouring scheme is also being considered within the same time frame and, if granted planning consent first, threatens the likelihood of the scheme gaining planning consent.	The active role of the local newspaper, particularly online, has previously fuelled misinformation about development projects.

While not all of these categories are necessarily relevant to a small planning consultation, they are worth bearing in mind to ensure an all-encompassing, comprehensive approach. It is no coincidence that the first section of this book, which sought to consider all factors affecting twenty-first century consultation, did so using the PEST template as its basis.

Box 14.2 Case study: initial research

Gunsko Communications/REG Windpower – Ivybridge, Devon

Gunsko was appointed by REG Windpower to run a public consultation on a commercial-scale onshore wind turbine proposal.

Stakeholder mapping was one of the first key components of putting together an effective communications, consultation and engagement strategy. It began with a detailed assessment of the site which looked at technical, environmental, political, planning and communications features and issues in relation to the site itself. Initially, the work was research-led and desk-based, and then investigated further in a site visit.

This approach helped the development team identify relevant stakeholder groups, by considering issues such as:

Technical:

- Are there properties/settlements along the access route that could be affected by construction traffic?
- Does the grid connection require any additional works, and if so, which individuals/properties could be affected by these?

Environmental (impact):

- What sort of environmental effects is the scheme likely to have, and whom would this be of interest to? (Local/national

environmental groups and statutory consultees, local residents, site neighbours etc.)
- Will the scheme be visible? If so, from where? (Properties, footpaths, local points of interest etc.)

Political considerations:

- Who are the political representatives relevant to the site/project stakeholders? (Parish/community councils, local/county councillors, MPs etc.)

Communications issues:

- What are the relevant local media? (Newspapers and online news, parish/local council newsletters etc.)
- What are the likely 'perceived' impacts of the schemes and how could this become a communications issue? Whom would this be of key interest to?
- What potential positive opportunities are there in relation to other local projects and aspirations?

Planning:

- Which is the relevant planning authority? (Planning team, planning committee members etc.)

This provisional list remained a live document and was supplemented as stakeholder engagement commenced. As a long-term local resident, the relevant landowner for the project was able to provide a wealth of local knowledge and suggestions, and further advice was sought from the scheme's planning officer. As stakeholders and communities became acquainted with the scheme, they were able to make further suggestions with regards to who else needs to be informed of the scheme and included in the consultation process.

Stakeholder analysis

Researching likely consultees enables the consultation strategy to take into account the appropriate number of residents to be targeted, the diversity within the community and an understanding of where power lies. Without it, a consultation runs the risk of being asymmetrical and failing to reach certain sections of the community, which often includes the 'silent majority' – those quietly accepting of the proposals.

A thorough understanding of the community also informs later stages of strategy development, for example, ensuring that tactics are well suited to specific groups. It enables the consultor to better understand the issues that motivate groups and individuals – such as something which may not be immediately apparent or may contradict initial perceptions. In a recent case, a political party supported a policy to bring about housing on a specific site, but despite being a member of that party, the local ward member personally opposed the proposal because of his fears for the repercussions on the local golf club of which he was a member. Party allegiances prevented him from speaking out in opposition to the scheme, but nevertheless he deliberately avoided constructive dialogue with the development team. A better understanding of the councillor's view at the early stages of the consultation would have enabled the team to communicate more effectively with the individual.

Stakeholder and political research tends to be interlinked, as local politicians are inevitably significant players in the local community. Similarly, communities of interest and communities of place co-exist (exacerbated by the increase in online communities), and most individuals fall into a number of categories.

Special interest groups are easy to identify, but thorough, ongoing, stakeholder research is necessary to recognise less formalised groups and patterns of interest. Comprehensive stakeholder and political research enables a much better understanding of those likely to take part in the consultation, and importantly, unearth useful and relevant insights. Typically the exercise will identify the following:

- The demographic profile of the area
- Local organisations – from community organisations to businesses and the issues affecting them
- Community/political/religious and special interest groups, their leadership, membership, policies and influences
- The political make-up of the council
- Planning committee members and ward members
- Political movers and shakers, including those with informal influence both within the council and in the wider community
- The likely impact of any upcoming elections on both the political make-up and individual roles
- Historic planning applications, particularly those for the site in question, or similar proposals which have been considered previously
- Anticipated or existing attitudes towards the development proposals among these stakeholders
- The history of local opinion towards proposed developments for the site, if any.

Box 14.3 Case study: stakeholder research

Atkins Global – Horizon Nuclear Plant, Anglesey

Horizon Nuclear Power, a UK energy company, is developing a new generation of nuclear power stations including the Wylfa Newydd project on the island on Anglesey in North Wales. Wylfa Newydd is a nationally significant infrastructure project (NSIP) and as such has very specific criteria regarding inclusion and reporting. The public consultation is run by design, engineering and project management consultancy Atkins Global, using a member of staff who works full time from Horizon's Anglesey offices and on site.

Consultation on Wylfa Newydd began with the publication of a community update newsletter, which was distributed to the 33,000 homes on the island and made available on the project website. The newsletter, which is published at each of the key stages of the consultation, sets out the path for the consultation and encourages those with an interest in the project to pass their details to the development team.

Atkins collates contact details on an ongoing basis following a variety of consultation events and continually manages and updates a database of names, addresses and other contact data. Customer relationship management (CRM) software is used because of its ability to produce statistics, share information via an app and create multiple tags which enables stakeholders to be analysed and grouped effectively.

Stakeholder and political research utilises the following:

- Office for National Statistics/Acorn consumer classification
- Demographic change – past and projected
- Town/parish council information and meeting records
- Neighbourhood forum information and meeting records
- Relevant planning documents and comments received by the local authority on previous applications in the vicinity
- Minutes of planning committee meetings
- Discussions with neighbours and 'key opinion formers' where possible
- Local newspapers
- Local blogs and community websites
- Social media
- Information about local groups (usually available from the local authority's community liaison officer)

Stakeholder mapping is a recent innovation which is proving extremely constructive in consultation.

The various methods of stakeholder mapping and research available have the potential to depict a community as an ecosystem, assessing the power and influence of individuals and gauging their likely reaction to certain issues. This enables the development team to understand a specific individual's likely view (be it positive or negative) and assess the influence of that view within the consultation.

Frequently, stakeholders are prioritised using a matrix. This approach plots power in one axis and interest in another (either may be substituted with location or any other relevant factor for measuring likely involvement), thus using the matrix to determine the prime focus for the consultation.

Box 14.4 Example political audit

Background to the council

- History
- Administrative structure
- Electoral history
- The political groups which make up the council
- Political considerations
- Economic considerations
- Social considerations (including ongoing public consultations)

The planning process

- The cabinet's role in planning
- The committee structure
- Planning officers
- Relevant ward members
- Members of the planning committee
- The council's Statement of Community Involvement
- Decision times and other statistical information
- Recent and forthcoming public inquiries

Planning background

- Local plan – its status; consultation to date; significant schemes proposed
- Area action plans
- Neighbourhood planning
- Community Right to Build initiatives and any other DCO applications

- Significant planning applications, both recently and those yet to be determined
- NSIPs affecting the locality

  Consultees and third-party groups

- Strategic partners – e.g. local enterprise partnerships
- Heritage and amenity groups
- Business groups
- Environmental groups
- Neighbourhood forums
- Parish councils
- MPs and MEPs
- Residents' groups

  Local press

  Local websites and social media profiles

  Conclusion: target stakeholders

At the end of the research process, the team will have an excellent understanding of the community generally, the personalities and groups which shape it and the issues which motivate or antagonise it. It should become clear which sections of society are likely to respond to the planning application. Information relating to key stakeholders, their contact details and relevant influences and opinions can be collated either in the form of an Excel database or held on customer relationship management software. Stakeholder engagement software is often used for larger projects as it has similar functions to CRM systems but can be customised to a specific project. It enables individuals to be pinpointed geographically and provides data distribution and analysis. The database should be continually developed, expanded and maintained throughout the consultation but used only for the purpose of the consultation and data protection rules adhered to.

Issues analysis

Having identified a broad range of factors impacting upon the consultation and those most affected, it is important to consider the specific issues that are likely to dominate the consultation. In communications theory, an issue is usually described as 'an unsettled subject ready for debate or discussion'.

In a local authority consultation, this would typically include the allocation of sensitive sites for development, and in development consultation, would include concern about the site, the proposals, or the impact of development more generally.

In some scenarios, typically in relation to strategic planning or larger, more contentious projects, a significant number of issues will already exist, some of which may be addressed by the consultation and others which are beyond its remit. To identify these issues and respond to them in a consistent manner, it is often advisable to compile an issues register at an early stage.

Knowledge of issues enables the development team to better understand the context of the consultation responses and, importantly, to address any misapprehensions. Inevitably, development consultations will involve some emotive and potentially divisive issues. Development on green fields, social housing and increased pressure on existing resources (roads, healthcare and education) frequently give rise to debate. Their impact, whether perceived or real, should not be overlooked, and the development team should respond to issues without delay, contradiction or confusion.

At the start of the consultation it pays dividends to put in place an internal issues or Frequently Asked Questions document. This sets out each of the issues likely to arise alongside the agreed response. The document must be flexible, as issues will change during the course of the project and themes will emerge or develop as new topics are discussed. Others may fall away as the community becomes reassured of the approach and misapprehensions are resolved.

Box 14.5 Identifying issues

The Concerto model – an intelligent community planning system by Woolley

In the mid-1990s, Nick Woolley chaired a steering group comprising the Environment Agency, English Nature and the RICS, which sought to develop a comprehensive means of using consultation to measure options for sustainable development on a range of projects and processes.

The result, which Nick then took forward, was Concerto. This used a pairwise system of analysis to prioritise issues. Sometimes known as paired difference and more commonly used in psychology, the system facilitates the comparison of entities in pairs, enabling the consultee to judge which entity is more significant when compared to every alternate option.

The three columns of sustainability – economy, environment, community – had been identified by Nick previously and are central to setting the agenda for a Concerto consultation.

The first stage of the process is the preliminary impact assessment. The development team invests considerable time in identifying potential issues within the three columns of sustainability. As many as 100 issues and sub-issues may be listed initially. These are then assessed for relevance on a scale of –3 to +3 and subsequently narrowed down to approximately 40 key issues.

A questionnaire is created which requires consultees to rank issues by number. Each issue is compared to another, with the consultee not only selecting the issue of greater significance, but also indicating the relative significance in percentage terms. Participants are also encouraged to add any additional issues later in the questionnaire. Thus the system is both quantitative and qualitative, and the process not only gains statistical responses to a set of issues which have been systematically developed, but allows local residents to contribute new issues too.

Purpose-built computer software provides a quick, clear and statistically representative results, which can be available within hours and then interrogated in various ways as most relevant to the project, with the issues set out in a graphic format that clearly highlights the key issues.

Concerto is unique to and copyright of Woolley and has been used for projects on behalf of local authorities, The Prince's Foundation for Building Community and private developers with notable success.

Box 14.6 Case study: Concerto in action

Woolley – major supermarket development, Falmouth

Woolley was appointed by a major food store to determine the appropriate positioning and detail of a new supermarket.

The process began in Falmouth, the preferred location, where local residents and other stakeholders were involved in issues analysis using the Concerto approach. Over 100 individuals were placed on tables of eight to ten people, with different interests and opinions deliberately spread throughout the room to provide the widest possible range of opinions at each table. A facilitator was present at each table to encourage discussion and make notes, always reporting consultees' comments verbatim. All those taking part were discouraged from challenging opinions at this early stage.

Following the initial discussions, each group collated the 12 issues that they felt were most significant, three in each of the four generic issue areas (environment, economy, social/community and infrastructure). These were then transferred to Post-It notes which were attached to a board for discussion with those of the other workshop groups. Each individual was then encouraged to rank the overall short-list of issues, and substantial discussion took place.

Following these initial consultation discussions, each participant then used the Concerto model questionnaire to identify and rank issues – both those identified by the development team at the start of the process and any additional issues that were felt relevant later in the questionnaire.

As is often the case, the day's activities were repeated at different times and over several days, due to the large number of local residents wishing to take part.

Following analysis of the results, Woolley produced a report which identified one new issue: the fact that affordable housing was seen as a particular concern in the local area and that without it, there would be significant staffing problems for the new supermarket. The client had been aware of this issue previously, but not to the extent that was made clear at the workshop.

The inclusive and open approach to issues management had been entirely instrumental in identifying not only the importance of the major problem issue, but also in identifying and prioritising all other positive and negative issues that were subsequently taken into account when the consultation entered the next phase.

Aims and objectives

The next and most important stage of strategy formation is to put in place aims and objectives.

Essentially, an aim is an overall aspiration for the consultation – for example, 'To engage with local residents as to their aspirations for the redevelopment of the town centre and, through working collaboratively, create a community-inspired master plan', or 'To consult all local residents and special interest groups on the proposals for Manor Hall Farm and to shape the development proposal, where appropriate, following material planning considerations raised during the consultation.'

Objectives should encompass everything that the consultation sets out to do by identifying the various specific, realistic outcomes to be achieved. It is crucially important that objectives are uncomplicated and easily measurable, as they will become the basis for the evaluation stage.

Box 14.7 Case study: creating objectives

Queensdale Properties Limited – Hamilton Drive, London

Dwyer Asset Management appointed PNPR to run a consultation on the development of a terrace of 10 townhouses, a large detached villa and landscaped gardens on some disused garages in St John's Wood, London.

The objectives for the development consultation were:

- **To inform:** ensure that all consultation materials contain a summary of the plans for Hamilton Drive, a description of the consultation process and contact details
- **To reach:** communicate the proposals for Hamilton Drive to all residents within a one mile radius
- **To involve:** engage with 25% local residents
- **Outcomes:** gain qualitative and quantitative feedback on design, access and landscaping and adapt the development proposals accordingly

---

Messages

The beauty of the strategic process is that each stage evolves from the previous stage. The messages section is no exception: it simply clarifies what has been identified in previous sections and must be communicated to consultees. The messages for Hamilton Drive were clearly linked to earlier stages in the consultation, as shown by the example below:

Table 14.4 Case study: using information to generate messages

Stage identified	Information gained		Message
Pre-consultation dialogue	Substantial pre-consultation dialogue with both residents and local councillors identified that those likely to take an interest in the development were those in the immediate geographical area, particularly those	→	Dwyer Asset Management, for Queensdale Properties Limited, is willing to meet interested parties on an individual basis, at their home or elsewhere in the neighbourhood, to discuss the plans in detail.

Stage identified	Information gained	Message
	overlooking or bordering the site. The demographics of the area suggested that those specific residents were likely to have a high degree of involvement and would appreciate a personal approach and in-depth discussions.	
PEST analysis	In replacing disused garages with high-value homes, the development was considered likely to increase house prices in the area.	→ The new development will benefit local economic prosperity.
SWOT analysis	Residents objected strongly and vociferously to the previous planning application.	→ Hamilton Drive differs substantially from the previous planning application: it is less dense, there are fewer units, affordable housing is provided off-site; the design is more in keeping with the neighbourhood; and the proposed underground car park is substantially reduced, resulting in little change to traffic patterns.
Stakeholder analysis	The stakeholder analysis identified specific individuals and groups most likely to engage and researched their power/influence.	→ We hope to hold constructive dialogues with the William Court Residents' Association, Grove Hall Court Residents' Association and the residents of Hamilton Gardens.
Issues	The question for residents of St John's Wood was not whether the development should be built or the land left as it was, but whether the new or consented scheme should be built.	→ The consultation will seek to determine whether residents support the new scheme over the consented scheme.
Objectives	Gain qualitative and quantitative feedback on design, access, landscaping and community facilities and adapt the development proposals accordingly.	→ We hope to discuss the following issues: feedback on design, access, landscaping and community facilities, to adapt the development proposals accordingly.

Messages must be clearly presented, consistent, easy to understand and communicated through a variety of media.

Strategic overview

Having set objectives and identified messages, the next stage is to put into place a strategic overview which sets out the overall approach and scope of the consultation. This will encompass a variety of factors, primarily revolving around *who, what* and *how.*

- What are the legal requirements?
- Who will be involved? Is it possible to involve the entire community or must the consultation be selective? If so, can this be justified? Will the selection be representative of the community at large?
- Is everyone relevant, or some more than others? What steps will be taken to voice the views of the hard to reach? It is important not to target the consultation towards those likely to support it, as the consultation would quite rightly be seen as biased.
- What are the subjects upon which a decision can be taken? Will consultees be presented with specific questions on limited aspects of the proposal, or are they invited to help formulate a master plan? Does the consultation welcome alternatives and new ideas, or simply request the local community's approval of an existing plan? Typically, local residents will be encouraged to comment on issues, such as access, community facilities, the construction programme, density, design, landscaping and additional uses, but other subjects may be available for discussion. In the case of NSIP consultations, consultation on community benefits is obligatory. It might also be worth considering consulting on non-material planning considerations. Asking local residents about the softer issues, such as what type of café or restaurant they might like to see included in a development, or the play equipment they would welcome in public areas, is often a good way of bringing people into the consultation process and can lead to comment on key issues, but is in itself not a material planning consideration and as such is not a necessary component of the consultation.
- Will the consultation allow people to respond anonymously? While anonymity has been shown to enable people to put forward their viewpoint without fear of repercussion, it could be argued that anonymous results cannot be verified at the evaluation stage and therefore carry little weight.
- What level of involvement is required? As described earlier, this depends upon the stage in the planning process and the options available. Many consultations will be formed of several stages, involving substantial engagement at the early stages and a narrowing of options towards the end.

Table 14.5 Considering anonymity in consultation

Arguments in favour	Arguments against
• Respondents are more likely to express their views without fear of repercussions • Breaks down power relations • Frees up individual expression • Removes bias • The argument can be focused on the content of the discussion without prejudice • Undermines collaboration	• A consultation report carries more value if comments can be attributed • An individual should be prepared to 'own' his/her comments • People are more likely to be dishonest when unidentifiable, or to use a forum to praise themselves • Individuals may be able to put forward their views on numerous occasions by using different log-ins • Anonymous contributions lack demographic data, which can be very valuable in a consultation • De-personalises comment

- Who will make the final decision? Where feasible, the developer should make a commitment to act on the results regardless of outcome, but in many instances although feedback is taken into account, a final decision is made by the professional team.
- What type of information will the consultation produce? For example, will the consultation seek results which are qualitative or quantitative, or both? Qualitative results are useful for exploring issues but are harder to evaluate; quantitative results provide a useful snapshot but lack depth and, to carry the necessary validity, require a substantial sample size.
- How will the consultation be carried out? This section of the strategy should specify, in very general terms, the diversity of tactics to be deployed and the timescales involved.

During a long-running consultation, the stakeholders, the context, and attitudes may change and the consultation strategy must be flexible to accommodate this change. This does not apply to NSIP consultations, however, which invest considerable time in pre-consultation research and consequently work to a fully formed consultation strategy from which, by law, the applicant may not be diverted.

External influences may also affect the strategy. For example, a local authority consulting on a local plan will need to take into account specific national planning policies, housing allocations and highways proposals; a property developer would be advised to check the local authority's SCI and hold discussions with its planners prior to determining the approach. Most companies will also have an overarching corporate strategy containing specific objectives and messages. Failure to take this into account could lead to inconsistencies.

The consultation mandate

The strategic direction of the consultation can be communicated with local residents through a consultation mandate: a distillation of the strategic overview for use by local residents. As this document will be read by a wide variety of people in a wide variety of circumstances; it is imperative that it is clear and concise, using plain language and a simple, accessible form.

Typically, a consultation mandate will include the following information:

- The organisation running the consultation
- The target audience
- The aims and objectives of the consultation
- The subject for discussion
- Potential impact of consultation
- The organisation initiating the change post-consultation
- Timings

Box 14.8 Case study: communicating the characteristics of a proposed development and consultation

Argent – King's Cross, London

Principles for a Human City was a document prepared by Argent (King's Cross) Ltd, the selected developer for King's Cross Central, and the landowners, London and Continental Railways and Excel in 2001 at the early stages of the regeneration of King's Cross.

The aim was to build a consensus on a shared set of aspirations for the regeneration of King's Cross before preparing detailed proposals. *Principles for a Human City* was an important part of that process. It set out the objective to devise and then deliver, over 15 years, an exciting and successful mixed-use development, one that would shape a dense, vibrant and distinctive urban quarter, bringing local benefits and make a lasting contribution to London.

Local stakeholders were invited to comment on *Principles for a Human City*, and the developer tested its ideas against these principles.

A robust urban framework

- Ultimately, the urban framework of routes and spaces is as important as the buildings and land uses which it serves.

A lasting new place

- Successful spatial master plans are long-lasting because they acknowledge and accommodate change

Promote accessibility

- Places should connect with each other, both physically and visually, and be easy to navigate

A vibrant mix of uses

- The development should have a vibrant mix of uses. This will help to create a place for people that is varied, enjoyable and generate lasting economic value.

Harness the value of heritage

- Heritage can contribute significantly to the sense of place necessary to generate economic, social and environmental value

Work for King's Cross, work for London

- The greatest destinations are products of their place, shaped by their local culture whilst addressing a worldwide audience

Commit to long-term success

- Sustaining the right blend of uses and activities requires ongoing, high-quality stewardship of the buildings and spaces, with an understanding of how each part influences the whole

Engage and inspire

- We will place particular emphasis on engaging actively with children and young people. They will live with – and hopefully benefit from – the redevelopment over the next 10 to 20 years. They may live and/or work there.

Secure delivery

- Our early development projects will be ambassadors for those that follow

Communicate clearly and openly

- We will create a clear, step-by-step process for this communication and engagement

Resource allocation

Timing

A time frame is essential. This is usually achieved by determining when the planning application is to be submitted and then scheduling pre-consultation planning and research, planning the consultation, collating and analysing the results, and amending the development proposal. Ideally, a consultation requires four weeks' pre-consultation planning and research, six weeks of consultation and a further four weeks to collate and analyse the results, making changes to the development proposal as required. It is advised that consultations are not held on bank holidays or during summer, Easter and Christmas and other religious holidays. Local or national elections are also best avoided, as planning issues could be taken out of context or their significance blown out of all proportion if they become the topic of a campaign.

A common complaint of developers' consultations is that they tend to be rushed. This is usually because an appropriate timescale is put in place but then slips due to lengthy land purchase, slower than expected appointment of a development team, and/or delays in scheduling pre-consultation discussions with local authority planners. Much of this is inevitable, but when it does occur should not result in a shortened time frame.

Human resources

A good consultation requires a good organiser, facilitator, strategist and processor of information – which may be present in one individual or several. In some cases, these characteristics, along with the necessary objectivity, are not contained within the development team and it may be prudent to put in place an individual to provide a bridge between the developer and a community. This role has been very effectively filled by a community worker, or sometimes an arts worker, who can quickly grasp the team's objectives but has knowledge of and links within the community and an ability and a remit which provides access.

Financial

Financial resources should be factored into a consultation strategy at an early stage. The benefits of a fully accessible, multi-faceted and all-inclusive consultation will invariably be reflected in its price. Budgeting for a consultation should take into account staff (including overtime, travel expenses and training); equipment, facilities and venue hire; advertising, PR, translations, document production and distribution; web design, development and hosting; and the collation and processing of information.

Creating an identity

A consultation benefits from a name and a visual identity which makes it clearly recognisable and separate from the identities of the partners involved. The new identity can sometimes be created as an early stage consultation tactic, thus providing residents with a greater sense of ownership. The new identity should aim to communicate the core ideology of the consultation, its values and purpose, perhaps providing subtle links through colours, visuals or fonts to partners.

Monitoring

Monitoring should not be confused with analysis or evaluation. The important difference is that monitoring occurs throughout the consultation. Analysis, although it can be ongoing, takes place (or is complete) at the end of the process. Likewise, although individual tactics can be evaluated while they are in progress, the consultation can only be fully evaluated when complete.

Unlike analysis and evaluation, monitoring need not be a formal, systematic process. Neither must monitoring be recorded. It is simply the process by which consultation tactics are observed. The benefits are two-fold: to ensure that tactics are working effectively and to enable the development team to take part in the dialogue as necessary. While the former is a necessary feature of all good consultations, the extent to which the latter is carried out varies considerably.

In some cases, the consultors' voice is rarely heard in the discussions; in others it may be necessary to clarify the messages and stimulate the dialogue to ensure an effective consultation. When determining to what extent monitoring should become involvement depends upon the following:

- Whether the consultation tactics are successfully meeting the consultation objectives or whether intervention would ensure greater success – for example, is the community well represented, or does work need to be done to bring others into the consultation?
- Whether dialogue is focused on the consultation objectives – is intervention required to bring the discussion back on track?
- The accuracy of the discussions – if misapprehensions have arisen it is usually necessary to provide clarification.
- Promises made to the consultees via the consultation mandate – did the consultation mandate stipulate that dialogue would be between residents, or between residents and the organisation running the consultation?
- The consultation's messages – are the messages receiving the necessary airtime, or does a particular message need to be brought to the fore?
- Bias – would intervention by the consultor be seen as 'leading' the results of the consultation?

- Symmetry and responsiveness – conversely, in not taking part in discussions, is the consultor failing to put across important information and to respond to points made?
- Information gathering – could more be learnt by asking questions?

NSIPs and the strategic approach

Chapter 13 described specific legal requirements in relation to NSIP consultations. It is worth addressing the necessary features of an NSIP consultation strategy here, however, not least because the stringent nature of the process necessitates a strategic approach which in many cases can be viewed as an exemplar for other planning applications.

The consultation strategy should aim to target the largest possible number of relevant stakeholders and in doing so be mindful of issues of accessibility, particularly in relation to hard-to-reach groups. When compiling the consultation strategy, the development team should be realistic about levels of involvement, making the process easy and efficient for busy people.

Table 14.6 NSIPs and the strategic process

Step	Strategy stage	NSIP requirements
1	Pre-consultation dialogue	The developer must put in place an appropriate approach of consultation in a Statement of Community Consultation (SoCC), with the local authorities responding on behalf of the general public.
2	Research	Initial research must be extensive. In addition to discussing the strategic approach within the development team, the applicant must consult external bodies on the process to be adopted. This includes all relevant local authorities, among others. Other informal consultation is advised to help shape the consultation.
3	Strategy mandate and objectives	The SoCC performs an important role in stating what can and cannot be influenced through the consultation. At this point specific objectives are put in place which, importantly, will provide the required standard of evaluation.
4	Tactics and timetable	The consultation timetable will be laid out in the SoCC and the applicant is expected not to alter the timetable. The timetable of the consultation will determine specific of consultation phases which will add clarity and structure to the process.
5	Feedback	The applicant must provide feedback to all consultees. At the very least, consultees should expect to receive a summary report (detailing the results of the consultation) and an invitation to attend the inquiry, should an inquiry be held.

Step	Strategy stage	NSIP requirements
6	Resulting actions following consultation	Applicants are required to show that due regard has been taken of all results, and specifically how the planning application was altered as a result of the consultation. Additionally, the SoCC must include, where appropriate, a statement showing why significant responses did not impact on the emerging development proposal.

Conclusion

Aside from any specific legal requirements, what constitutes an effective consultation strategy? Regardless of the size, scope, timing and extent of the consultation, a good strategy will deliver:

- The clearest possible understanding of the background to the scheme and the issues impacting both on the development proposal and the consultation itself.
- Comprehensive information about the target audience, including contact data and intelligence which may benefit the likelihood of reaching specific individuals/groups.
- A clear direction, to be understood and shared by the team (developer, architect, planning consultant, public affairs consultant) and external bodies such as local authorities, government agencies and local enterprise partnerships.
- Consistency of voice among team members, ensuring that the nature of the consultation, its purpose but also its limitations, is communicated effectively; likewise the team's commitment to the consultation.
- The purpose, approach and timing of the consultation, outlined in a single document to avoid confusion.
- A means by which the results can be collated, analysed and used.

This chapter has addressed the need for a consultation strategy in relation to communication, planning and resourcing. It has shown that consultations can vary significantly in terms of their strategic objectives, but that a standard strategic framework exists which can provide an excellent basis for a consultation regardless of the remit of the project, the legal requirements or the resources available.

Taking into account the issues addressed, we conclude with a set of guiding principles which may be used to ensure quality in consultation.

Principles of good consultation

Strategic

A good consultation must be well researched, based on firm objectives, structured and designed to produce meaningful analysis and evaluation.

The strategy should be well understood within the development team and communicated to wider audiences in the form of a consultation mandate.

Two-way

A good consultation aims to achieve a symmetrical flow of information between the development team and the local community, as opposed to bombarding the community with information and paying little attention to responses.

Responsive

The day of informing the public on a development proposal and collating results at the end of the process is over. Consultations focused on ongoing involvement allow development proposals to evolve in line with feedback, and allow the process to adapt where necessary.

Positive sentiment in a local community brings about positive consultation responses. An ability to understand what motivates a community and to communicate with local residents in the way in which they feel most comfortable is crucial.

Box 14.9 Case study: ongoing communication with the community

Gunsko Communications/BNRG Renewables – New Barn Farm, West Sussex

Gunsko was appointed by BNRG Renewables to run a public consultation on a commercial-scale ground-mounted solar PV development. The consultation required that all local stakeholders were given an opportunity to respond to the proposals, and therefore it was necessary to throw as wide a net as possible to capture all levels and types of stakeholder that might have an interest or be affected. This included local environmental interests, representatives of the local community, elected officials, site neighbours and the local planning authorities.

Stakeholders were contacted in the earliest stages of development, six months ahead of a formal planning application being submitted. This was important to allow sufficient time for genuine consultation to take place.

Gunsko kept stakeholders updated on project progress using relevant media, including a bespoke project website, mailshots, presentations and meetings, and operational solar farm visits.

Incorporating a long time frame for consultation allowed stake-holders, communities and the development team the opportunity to truly engage with each other, tease out genuine issues of concern and mitigation options, identify and incorporate positive opportunities and dispel misapprehensions. As a result, local stakeholders were able to provide constructive feedback to the proposals, feeding into an evolved design.

Genuine

Honesty is at the root of all good consultations. Transparency and openness should be present throughout, from setting realistic objectives, communicating the purpose of the consultation, drawing up agendas for discussion and reporting events to the final feedback. The consultation should avoid any association with spin.

Information should be shared openly, especially documents which explain the proposals in detail and are therefore likely to impact on the final result.

Similarly, a consultation outcome should never be pre-determined. Often, factors will come to light which do not support the final decision. The consultation report can identify such issues and explain if necessary why they were not represented in the final decision.

Engaging

A positive approach is imperative: a good outcome is the result of a good process. An engaging consultation might mean fun to some stakeholders, inspiration to others, regular and consistent communication to others. The consultation strategy must be mindful of the various groups that make up the community and seek to provide relevant and appealing forms of engagement.

A successful consultation is one which results in constructive relationships with the local community. Formal partnerships, such as with potential occupiers of community buildings or with local enterprise groups, are often of substantial benefit, both during the consultation and during the construction phase.

Consultation should be clear on every level, from the language used to the communication of the aims and objectives. A consultation report provides an ideal opportunity to clarify those consulted, the information gained and the impact upon the final scheme.

Timely

Early engagement is good engagement. A good consultation allows ample time to develop the early stages of the strategy, to engage fully and provide adequate time for responses. The timescale of the consultation should be set out in a document which can be accessed by all.

Box 14.10 Case study: early stage consultation

Terence O'Rourke Ltd – Milton Park, Oxfordshire

A former MOD site near Didcot, Oxfordshire, Milton Park is a 100-hectare business park owned and operated by MEPC and is home to several hundred businesses. In 2012, MEPC and its planning consultant, Terence O'Rourke Ltd, worked in partnership with the Vale of White Horse District Council to introduce a simplified planning regime for the park. The Milton Park local development order (LDO) grants advance permission for specified land uses provided they meet prescribed development parameters.

The idea of rolling back planning controls, including the loss of statutory public consultation associated with individual planning applications, was initially met with concern.

Terence O'Rourke therefore invested in pre-consultation dialogue prior to the preparation of the LDO. This focused on early engagement with local politicians and the wider community to explain the purpose and merits of the LDO and foster involvement in the emerging development parameters. An open community consultation was held at the park to raise awareness of the initiative and to invite comments on the key planning issues. This exercise was successful in raising awareness of the proposed LDO, engendering trust and addressing initial concerns. It led to direct involvement of the parish councils, with a special community meeting to discuss the proposals.

As well as contributing to the detail of the development parameters, a key outcome of this early engagement was the establishment of a community liaison group. The LDO was adopted in 2012, but the group still meets regularly and continues to provide a channel of communication between MEPC, the councils and parish councils. The community liaison group is chaired by a local councillor and provides the forum for MEPC to notify the community of forthcoming developments, understand and address concerns that arise and monitor the success of the scheme.

Informative

A consultation is a learning journey: from informative research to a comprehensive and a well-reasoned argument in favour of a scheme in the form of a consultation report. The development team should see its role as one of learning about the community from the community, and this approach should be apparent in its dealings with local people.

**Box 14.11 Case study: the benefits
of anecdotal information**

Peter Brett Associates – Collingtree, Northamptonshire

Peter Brett Associates (PBA) worked with Bovis Homes on a new development of 1,000 homes, a mixed-use local centre, a primary school and public open space in South Northamptonshire. The proposal also involved the reconfiguration and extension of Collingtree Park Golf Course.

PBA carried out substantial consultation, both formal and informal. It was through informal discussions with local residents and users of the golf course that the development team became aware of the potential for flooding on the site, including some very specific instances of where and how flooding had occurred previously.

Although a flood risk assessment was already underway, this anecdotal information enabled the development team to focus its assessment and helped inform the flood management strategy. As a result of both professional and local input, the master plan was adapted in such a way that water flows were redirected to reduce flood risk.

Managing expectations

Expectations within a community are likely to be as varied as the individuals that constitute it. This is best understood through thorough comprehensive stakeholder research, not only of information but of feelings and expectations. With a better understanding of local people's expectations, both for the consultation and the scheme, the developer is better placed to communicate the parameters and prevent disappointment. Care should be taken to balance the need to motivate residents to secure their involvement and the need to set realistic expectations.

Box 14.12 Case study: correcting misapprehensions

**Land Use Consultants (LUC)/RWE Innogy Ltd – River
Valley Wind Farm, Yorkshire**

Research has shown that people are generally in favour of renewable energy. However, developing sites for onshore wind is often controversial and can lead to opposition from local residents. Although developers may not change the opinion of local residents, it is important that they work with the community to ensure that the potential effects of the scheme are fully understood. By achieving this, the discussions can focus on the real issues rather than misapprehensions.

LUC worked with RWE Innogy Ltd on their proposed six-turbine River Valley Wind Farm in the East Riding of Yorkshire. The local parish council was vocal in its opposition to the scheme, which both RWE and LUC recognised was partly due to misunderstanding over the scale of the scheme proposed. The land area required for a wind farm can often be much larger than the development itself, especially during the design stage when the position of infrastructure is not fixed. It was clear from conversations with the parish council that they believed that RWE had intentions to extend the scheme in the future, on to land shown as within the development boundary but distant from the current proposal. This was far from the case.

Land is often constrained by technical and environmental factors such as proximity to residential properties, while some areas may be included within the development boundary for biodiversity enhancement purposes. Recognising the concerns of the local community, RWE and LUC took the decision to hold a workshop with the parish council to explain the design process. The level of detail and information shared was uncommon in comparison to other developments, but the feedback from the parish council members was constructive. Following the workshop, they understood that sections of the site were constrained by buffers required around ecological and hydrological features, residential properties and telecommunications links, and that areas to the north were allocated for biodiversity enhancements.

While the parish council remained opposed to the scheme, having taken them through the design process, LUC was able to move the discussions on, generate a good working relationship and alleviate their concerns regarding a potential future extension. This process built a level of trust between the parties and enabled further productive consultation on the remaining key issues.

Understanding social value

The process of consultation always serves as a reminder that development is much more than physical buildings and infrastructure. Size, massing, design and landscaping are only part of the picture as far as local residents are concerned. Development is a backdrop to life, and individuals' priorities in life differ hugely. Community integration, a social scene, culture and heritage, access to facilities and house prices are just some of the many ways in which communities are defined by those who live in them, and these factors are likely to shape consultation responses.

Accessible

The visibility of those carrying out a consultation is a strong indicator of its likely success. Developers, however senior, should be seen to listen and learn.

Box 14.13 Case study: an accessible figurehead

Argent – King's Cross

Roger Madelin, former CEO of Argent, is widely believed to set new standards in making a consultation accessible – primarily in his personal approach to community engagement.

Madelin began the consultation by promising to speak to local residents 'anywhere, anytime' and during the four years of consultation the commitment was upheld, no fewer than 7,500 people were met in a variety of consultation events and meetings in two years alone. An accessible, non-corporate approach was key to this success. Madelin integrated at all levels, talking to older residents over tea and playing football with the young.

Madelin firmly believes that local residents are the best source of information and inspiration and are the best means of enabling a new development. Many of the decisions that were taken at King's Cross, particularly in relation to maintaining the heritage and character of the area, would not have come about without substantial local involvement and have considerably benefited the character of the development.

Note

1 A consultation report is also commonly known as a Statement of Community Involvement (SCI), but to avoid confusion with the consultation requirements put in place by a local authority, also known as a Statement of Community Involvement, the term *consultation report* will be used throughout this book to refer to the document which reports on and analyses consultation results and forms part of a planning application.

15 Tactics to inform and engage

Introduction

The previous chapter explained the stages of consultation strategy development – understanding the neighbourhood and the issues which motivate it (situational analysis, stakeholder research and issues analysis) and putting in place the structure of a consultation (creating aims and objectives and formulating a strategy and timetable). Chapters 15 and 16 address the next and most substantial stage: how to inform people of the potential for change and to consult effectively.

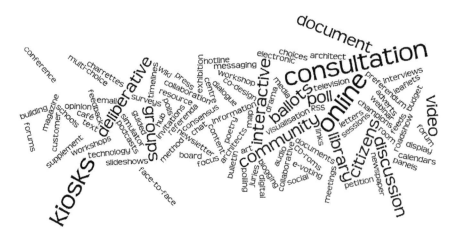

Image 15.1 Word cloud showing relative popularity of consultation tactics

Informing and consulting are very different tactics. As Chapter 14 demonstrated, good consultation is two-way. The process of informing an audience however, is primarily one-way. This is shown in Figure 15.1.

In addition to those which exist solely to provide information, there are two further categories of consultation tactics: those which deliver primarily quantitative results and those which deliver primarily qualitative results.

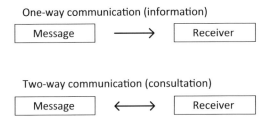

Figure 15.1 One- and two-way communication

Information tactics

Imparting information typically includes use of existing print and broadcast media, letters, newsletters, provision of information online, presentations to large groups, email and text messages.

Most commonly, information tactics are used when a development proposal is already in place and it is necessary to inform local residents of its content prior to consulting. Thus the community has an understanding of the proposals at the point at which they are invited to comment, and responses are more meaningful as a result. In an NSIP planning application, proposals are often progressed to a considerable extent before the community becomes involved.

Providing information is a helpful means of addressing any misapprehensions at an early stage. Information tactics are also an effective means of raising awareness of a consultation in its early stages and can be used to target a wide range of potential consultees. However, information tactics are all too often heavily relied upon as the main means of communication. Furthermore, a 'scattergun' approach risks individuals receiving information that is irrelevant to them or that they do not understand.

It is important, of course, to ensure that the number of information-only tactics are limited. A consultation which informs local residents about proposed changes but gives no opportunity for local residents to make views known or ask questions is asymmetrical and has potential to generate anger and mistrust.

The one-way nature of these tactics allows the consultor to control the output – promoting key messages, correcting known misapprehensions and providing a well-rounded image of the organisation through branding and corporate image. But such tactics do not offer consultation *per se* and, in some cases, it is worth considering potential to convert an information tactic to a consultation tactic simply by providing a response mechanism. This can often be done without significant effort or expense and is worth considering, providing the information received can be analysed effectively and that the use of the consultation tactic fits with the overall objectives for the consultation.

Quantitative tactics

Tactics which produce only quantitative data tend to be those that target a wide audience which, it is assumed, has some prior knowledge of the scheme. Typically, therefore, these tactics are used at the end of the consultation process to gain a clear view of local opinion. An example of this is referendums, which are increasingly used due to neighbourhood planning and the new Community Rights. Other tactics which deliver quantitative results include polls, questionnaires and surveys.

Previously, quantitative techniques were widely used because of their ability to produce large quantities of definitive data quickly. However, their use has waned as more collaborative tactics, such as Planning for Real®, Dialogue by Design and Enquiry by Design,[1] have been introduced and have exposed the difficulty in generating meaningful responses from quantitative tactics. That said, the internet has introduced many quick, free and simple means of polling, resulting in data being collected and analysed considerably more easily than previously. There have also been extensions to traditional methods which introduce a more qualitative element. The preferendum expands the concept of a referendum by including a selection of options in place of yes/no questions. Similarly, deliberative polling significantly increases the levels of feedback possible through a standard poll combining public opinion research and public deliberation to construct hypothetical representations.

The value of quantitative data relies both on a scientifically significant sample size and on information being readily available to those participating. A well-planned approach focused on objectives can provide some very useful responses, whereas tactics which fail to ask the right questions of the right people and produce results which are not a true reflection of opinion can be harmful to the consultation.

Qualitative tactics

Qualitative techniques are those which bring about an opinion or observation (in words) rather than a statistical result (in numbers). Qualitative tactics are increasing in popularity with the rise of participatory planning and also the internet, which can facilitate excellent dialogue.

Online, qualitative tactics include blogging, chat rooms, forums, emails, and interactive displays, layering and adapting maps, discussion groups, tech tapestries, video soapboxes and virtual meetings. Offline, they might include social initiatives such as a consultation cafés and art/theatre workshops; those techniques based around dialogue and discussion (in-depth interviews, discussion forums, discussion group, face-to-face interviews and focus groups); collaborative planning techniques such as charrettes, citizens' juries, planning workshops and policy conferences; the setting up of task forces and think tanks; and opportunities to find out more about the scheme which then result in dialogue – such as exhibitions, roadshows and pop-up events, site visits and seminars.

Box 15.1 Case study: using models in consultation

David Lock Associates/Yorkshire Forward – Halifax Urban Renaissance

The Renaissance Towns project was developed by Yorkshire Forward, the regional development agency for Yorkshire and the Humber. Halifax was one of the first towns selected to take part in the project.

David Lock Associates (DLA) was appointed to work with the Halifax Town Team, a group of key stakeholders who had a vested interest in the future of the town. The aim was to create a vision and development framework for the town centre for implementation over a 25-year period. The Town Team was critical to driving change and becoming 'civic leaders'.

The focus of the project was to improve the quality and connectivity of the public realm. The town centre has many fine assets or 'crown jewels', but these were poorly integrated. So the engagement programme began with a series of 'town trails' which examined the key approaches into the town centre, how to improve them and better connect the 'jewels' of Halifax.

An important component was a week-long children's workshop, held at the Eureka children's museum, which involved 40 children from three local schools building a large-scale model that reflected their ideas and aspirations for the future of Halifax.

The event began with a presentation which described the ambition, issues, context and limitations. The children, working in groups, were then encouraged to create an alternative neighbourhood using models and mosaics. Representatives from each group then explained the model to the wider group and a discussion followed.

DLA used the ideas to formulate a master plan, presented the plan to local residents and gained their approval.

The ideas were incorporated into investment funding applications and a non-statutory strategic development framework plan, *Halifax – Streets Ahead*. Concepts and objectives were adopted by the Town Team and used to inform their strategy for action and intervention across the town.

Those involved were in agreement that the workshop was successful in bringing about ideas that may not have been generated otherwise and reaching those who might not otherwise get involved.

Box 15.2 Case study: online forums

Polity/ConsultOnline – Next Home and Garden Store in Handforth Dean, Cheshire

In its comprehensive consultation on behalf of Next plc, Polity worked with ConsultOnline to create an engaging online strategy, specifically designed to appeal to younger people and the large digital community which already followed Next on Facebook.

Over a period of six months, the Facebook page communicated with local residents through news updates, polls and discussion forums. The latter was the most successful method of engagement, with substantial numbers of people discussing issues such as local employment, features of the new store, whether a café was required as part of the proposals, the design of the store, changing shopping habits, and responses to the consultation itself.

Not only did the online forums provide information which was invaluable when drafting the consultation report, but subjects were shared via social media, thus increasing the numbers of those involved in the consultation. Furthermore, the use of 'soft' subjects to engage people initially enabled the development team to target the same respondents with other questions relevant to the consultation.

Qualitative tactics compliment quantitative tactics, not only in the results generated but in the approach taken. Due to the emphasis on dialogue, qualitative tactics provide an opportunity for those running consultations to develop constructive relationships with key stakeholders.

Qualitative techniques have the potential to gain significant insight into a particular topic, with a good facilitator able to probe further to unearth issues. Tactics such as focus groups are particularly useful in ascertaining thoughts on specific subjects, while a series of online forums or a day-long workshop can consider a range of issues. Qualitative techniques provide detailed feedback, unearthing feelings as well as facts. Furthermore, discussion can be effective in sparking ideas and thoughts – although the reverse is also true as gossip can result in the fears escalating if messages are not clearly communicated and dialogue monitored.

A facilitator is therefore vital for the successful execution of discussion groups, whether the discussion is conducted on- or offline. A facilitator requires a clear agenda, but must avoid steering responses in a particular direction. Listening and reporting also requires significant skill: data is transient until recorded and all too often excellent feedback is lost because the tasks of facilitating and recording group conversation cannot be carried out simultaneously.

It is important to remember that results generated by qualitative tactics are not necessarily representative of the wider target audience. For example, activities which involve a considerable time commitment are likely to attract those in the 65–75 age group; conversely, online consultation is more popular among younger people, although the 'digital divide' is closing. If using pre-existing groups rather than focus groups, it is important to consider that representative organisations have a particular interest or strong views on a subject. A stakeholder, or project liaison, group set up for the specific purpose of the consultation can address these factors, but even so, it is very difficult to create a group which is genuinely representative of the wider population.

Box 15.3 Case study: creating a project liaison group

Atkins Global – Horizon Nuclear Plant, Anglesey

In its substantial consultation on the Horizon Nuclear Power project (Wylfa Newydd) on the island on Anglesey in North Wales, Atkins Global set up a project liaison group.

The purpose of a project liaison, or stakeholder, group is to develop strong links with local residents, representatives of community groups and in some cases local authority members and officers, with a view to imparting key information and gaining feedback on a variety of issues.

Approximately 100 individuals are invited to attend Atkins's quarterly project liaison group meetings, and typically 60–80 will attend, of which 40 are regular attendees.

Each project liaison group meeting begins with an update. This may include anything from local consultation events to key developments with external bodies, such as the National Grid. A specific topic of conversation is set for each meeting and following discussion on this point, Atkins gives those present an opportunity to ask questions of the project team.

The project liaison group has proven to be an excellent means of developing and maintaining relationships with local residents, gaining important insight and working with local people to suit the new power plant to the community.

Communication through dialogue is an excellent opportunity to build bridges into a local community and should not stop with the event itself, instead providing a new network of contacts and the opportunity for ongoing dialogue.

Building in analysis and evaluation

At the planning stage of a consultation, it is crucial to determine how data will be generated and later analysed: a two-way tactic which delivers unusable information is essentially a one-way tactic. For example, a leaflet is drafted as the first stage of a consultation to inform local residents about a supermarket being built on the outskirts of the town. Mindful of the need for two-way communication, the developer includes a freepost comment form on the back of the leaflet and 500 responses are received. Without systems to collate and understand the responses and use them to inform the issues database, this time and expense is not only wasted: when local residents hear nothing in response they invariably grow resentful of the consultation and angry petitions and protests take the place of meaningful dialogue.

Similarly, it is very important that the tactics are designed in such a way that they can be evaluated.

Asking the right questions

What are the 'right' questions to ask in a consultation? Unsurprisingly, there is no 'right' answer! As has already been identified, consultation ranges from an issues-based exercise which encourages a wide range of ideas from its audience to a referendum which invites residents to vote 'yes' or 'no' to a single idea. Ideally, a consultation will include both open and closed questions.

Depending on the nature of the consultation, fully open questions such as 'What do you think the development should comprise?' can give rise to unrealistic answers or angry rants. It is often more helpful, both for the individual and the exercise, to provide guidance which focuses the mind and in doing so generates more meaningful responses. Issues-based consultations sometimes use the 'dilemma' approach: one that puts the consultee in the position of the consultor and in doing so helps them to make a more informed choice. For example, rather than asking the question 'What do you want to see on this land?' the dilemma approach would state, 'We are required to provide between 800–1,000 homes and three commercial units on this site. Where do you feel the commercial units could be situated? What sized homes are most needed in the neighbourhood? Do you agree that the 30% housing association homes should be distributed evenly throughout the development?' The alternative approach – seeing the proposal from local residents' point of view – can also help address underlying negativity. For example, a telecoms company invariably faced with the comment, 'I don't want that mobile phone mast obscuring my view', may ask questions relating to *need* at an early stage, starting with the question, 'Is your mobile phone coverage satisfactory?'

Should a consultation request demographic data? Most consultations will benefit from a profile of their respondents. In a site-specific development, it is extremely useful to understand where people live and take this into account in relation to their response. Information relating to age, gender and employment status can also benefit analysis, but can be off-putting. Rarely is it worth asking for demographic information if it deters a significant number of potential respondents. A tried and tested technique is to seek this information at the end of the process, rather than early on: impart information, encourage feedback, acknowledge the contribution, and only then ask for demographic data to add additional value to the consultation response. It is also advised to make the provision of personal information voluntary, while both explaining its benefits, reassuring respondents that the information will remain confidential and not used for any other purpose. The Information Commissioner's Office provides useful information about handling personal data, and anyone running a consultation should consider registering under the 1998 Data Protection Act.

While data which lacks user information lacks validity, an anonymous contribution is usually more valid than none. Online consultation demonstrates that anonymity can benefit a consultation in removing hierarchies. In an online consultation conducted by ConsultOnline,[2] 54% of those taking part in the consultation chose a username which bore no resemblance to their actual name, yet names, addresses and postcodes were supplied for the registration process. The lack of these comments would have been detrimental to the consultation, and while respondents were reassured that their personal details would not be made public, the development team had access to the demographic data necessary to create an excellent consultation report.

Selecting tactics

As has been shown, a plentiful supply of consultation tactics is available. While it is necessary to use a variety of techniques to elicit a response from a range of individuals and groups, development teams are advised not to undertake so many tactics that the results become confused. The scope of the consultation will determine how many different tactics are appropriate.

Figure 15.2 shows a sample of consultation tactics in relation to the results they generate, the extent to which they engage with the audience and their approximate cost.

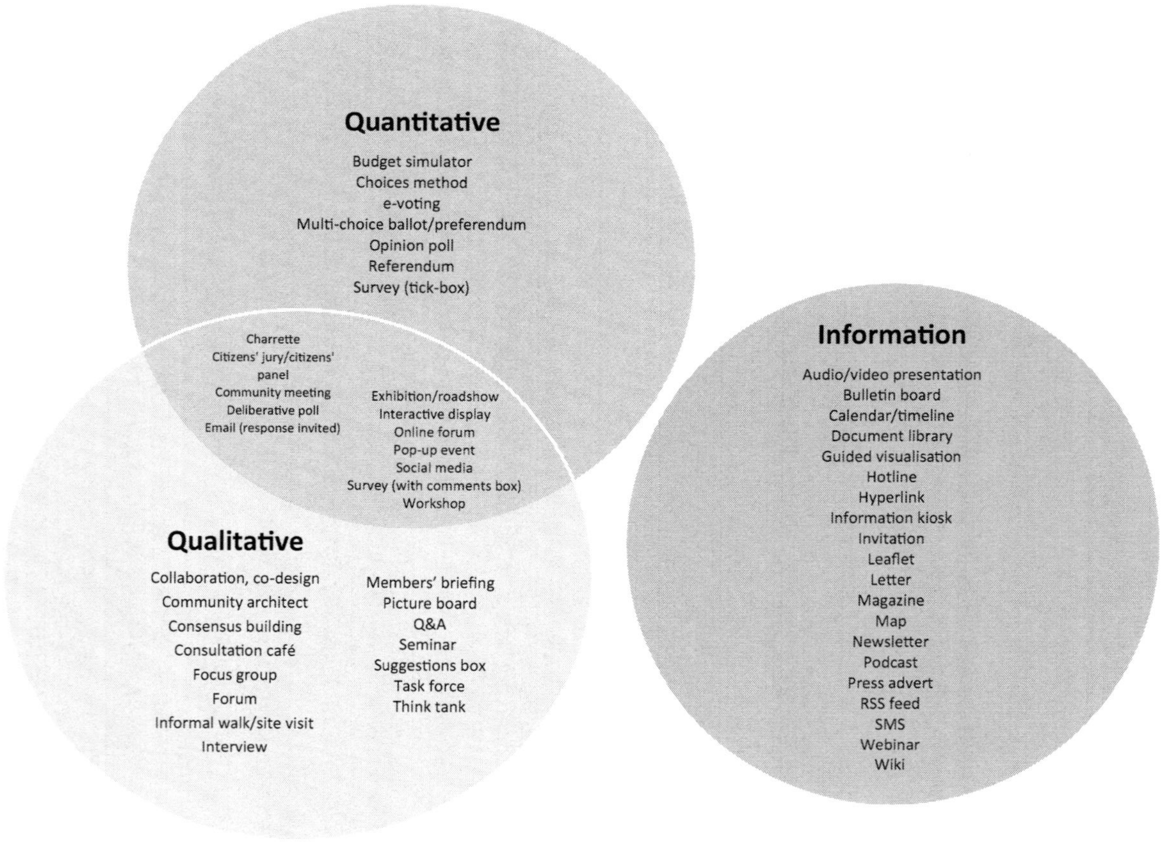

Information

Audio/video presentation
Bulletin board
Calendar/timeline
Document library
Guided visualisation
Hotline
Hyperlink
Information kiosk
Invitation
Leaflet
Letter
Magazine
Map
Newsletter
Podcast
Press advert
RSS feed
SMS
Webinar
Wiki

Quantitative

Budget simulator
Choices method
e-voting
Multi-choice ballot/preferendum
Opinion poll
Referendum
Survey (tick-box)

Charrette
Citizens' jury/citizens'
panel
Community meeting
Deliberative poll
Email (response invited)

Exhibition/roadshow
Interactive display
Online forum
Pop-up event
Social media
Survey (with comments box)
Workshop

Qualitative

Collaboration, co-design
Community architect
Consensus building
Consultation café
Focus group
Forum
Informal walk/site visit
Interview

Members' briefing
Picture board
Q&A
Seminar
Suggestions box
Task force
Think tank

Figure 15.2 The nature of consultation tactics

In selecting tactics, the following considerations should be taken into account:

- Accessibility – do the tactics selected give all sections of the community an opportunity to comment?
- Analysis – consider the outputs required for a convincing consultation report, including achieving a balance of qualitative and quantitative responses
- Anonymity – consider the benefits and drawbacks in relation to the consultation's objectives
- Appeal – make it fun

Box 15.4 Case study: making consultation fun

Atkins Global – Horizon Nuclear plant, Anglesey

The development team at Horizon Nuclear Power was aware of the need to make the consultation on the Wylfa Newydd project genuinely engaging, particularly the need to appeal to younger people.

A decision was taken to introduce the key topics for discussion as a game through which participants were able to learn about the new power station and the work that Horizon was involved in throughout the island.

As part of the game, each participant was given a token and asked to use it in support of the topic of greatest important. Topics covered housing and job opportunities but also subjects specific to the projects, such as the impact on the Welsh culture and language.

The participants responded very positively to the initiative, so much so that they requested that Horizon repeat the event on a future occasion.

Box 15.5 Case study: pop-up events

Transport for London – Bank Station capacity upgrade

The consultation surrounding the upgrade of Bank Station was complicated by its location in the historic City of London, involving conservation areas, world-famous Grade 1 listed buildings and the headquarters of major corporations.

It was vital to form collaborative relationships with local stakeholders and bring about constructive dialogue. In doing so, it was considered important to meet people on their own territory. Consequently, Transport for London (TfL) held a four-day exhibition in the seventeenth-century St Mary Abchurch, a church impacted by the proposals.

St Mary Abchurch opens onto a small but popular square which features bars and restaurants. The church itself, however, presents a blank face to the outside world, with a solid wooden door and high, frosted windows. Furthermore, access to the church was compromised by several steps leading to the main entrance.

Although the church was an ideal location, activity inside would not be immediately visible to those passing. So the consultation team installed a marquee, together with flags, outside the front door. This was successful in attracting attention and provided a welcoming reception. Any consultees who needed a fully accessible route could then be accompanied from the marquee through a side door into the church. The church had historic box pews, upon which display boards were mounted. Specifically, by attaching exhibition materials to the pews without fixings, the team demonstrated, albeit in a small way, the respect it intended to pay to others' property.

The focus of the exhibition was the model of the proposals which showed the complexity of the station in three dimensions. It was accompanied by an interactive 'fly-through', enabling visitors to navigate their way around the virtual station. The popularity and effectiveness of the fly-through was in inverse proportion to age, with the most enthusiastic participant being a 15-year-old boy who managed to navigate his way outside the station altogether and view it from outer space, to the amusement of other consultees!

The presence of several approachable, well-informed staff and planning/construction/transport specialists significantly benefitted the consultation.

The exhibition attracted 600 visitors over four days, forming an integral part of the wider consultation. Many people left having made contact, swapped business cards or arranged follow-up meetings. Due to the nature of the planning consent, the full extent of consultation activity was reviewed by Queen's Counsel and was found to be fit for purpose. The public inquiry took place 18 months later. There were no objectors.

- Balance innovation and more established methods
- Cost – do the chosen tactics fall within the consultation budget?
- Ease – avoid requesting unnecessary information or making it difficult for individuals to respond

- Mix old and new means of communication to appeal to the various demographic groups within the community
- Past successes – consider what has worked well in the past, or discuss successful local consultations with the local authority and local groups
- Time – assume no prior knowledge; give people time to digest information
- Variety – don't rely on just one method: different tactics appeal to different people

Most importantly, the strategic framework as described in Chapter 14 is the first way to ensure that the most appropriate tactics are selected: research of stakeholders, situations, and issues; identification of objectives; and a consideration of practical viability enable the development team to determine the means of consultation.

Box 15.6 Case study: collaborative consultation

North West Cambridge

The single largest development in Cambridge University's 800-year history covers a 150-hectare site. Outline planning permission was granted in 2013 to provide 3,000 homes (half for qualifying key university workers), 2,000 student bedspaces, 100,000 sq m of research space, as well as community facilities and public realm to ensure the long-term growth for the university and the city.

As a strategic growth site in Cambridge on the former green belt, with the university as principal developer, the challenge was to ensure that the new development would be supported by the existing local community through planning and into the long term as a new part of the city. The approach to consultation was taken with significant investment in building relationships with the strong local communities, which began in 2009 following the appointment of an experienced project director.

The consultation programme included public exhibitions, mailings, and focus groups. A dedicated website with e-marketing capability provided a vital communications platform to provide news updates and handle enquiries professionally yet individually.

Comprehensive mapping of all stakeholders within the local community was undertaken. Through individual meetings with residents' groups, community leaders and specific interest groups, meaningful dialogue and two-way communication was established. The university formed a consultative community group, which brought together community leaders to discuss proposals as they emerged in advance of the submission of the planning application. The consultation was initiated with a series of charrettes or stakeholder workshops which enabled residents and communities to be engaged with the technical team at an early stage, to inform the proposals and look at the constraints and opportunities of the site. Feedback inspired design features and led to various commitments being written into the development's design code. Throughout the four-year period of consultation, extensive discussion and consultation was undertaken with the local authorities and communities, which was praised by the former as collaborative, not just consultative.

Engaging communities with collaborative and participatory consultation has resulted in high levels of positive engagement among the local communities. The outline planning application was unanimously approved and the shared vision for the development is understood by the local communities, along with positive anticipation for the new development.

Notes

1 These forms of consultation are described in more detail in Chapter 16.
2 See the Scotch Corner Designer Village case study in Chapter 16.

16 New consultation tactics

Introduction

Three major factors summarise the changes that have taken place within consultation since the new millennium: the demise of the public meeting, the rise of participative planning and the introduction of online consultation.

The demise of the public meeting and the rise of participatory planning

'A lot of heat but no light' is an apt description of the majority of public meetings. In the 1990s, an open meeting at a local village hall or school was regarded a necessary element of most consultations. The development team would prepare a substantial amount of information and aim to talk about the proposed scheme long enough to minimise comment, mindful that a malcontent was invariably present, ready to jump up, oppose the scheme at the first opportunity and rally neighbours into an angry frenzy. Local residents, having received a letter or newsletter inviting them to attend, would do so only if they had a strong point to make: otherwise what was the point in venturing out on a wet December evening to sit in a draughty village hall? The opposition invariably got their point across, often amplified, whereas the views of the 'silent majority' would stay at home because they were happy for the proposals to go ahead, but consequently their acquiescence/support would not feature in the consultation responses. The local media in search of a good story and the local councillors in search an issue to boost their local support would attend in anticipation of a dramatic evening.

Not all public meetings took this form, of course: many were very constructive and continue today in some circumstances. But generally, development teams have sought more engaging and constructive means of consultation.

Participatory planning – also referred to as community planning, community visioning or collaborative planning – is gaining increasing prominence, and its variety of different forms is providing a welcome alternative to the beleaguered public meeting. Enquiry by Design, Planning for Real® and JTP's community planning weekends are just a few examples of this growing phenomenon.

Box 16.1 Case study: intelligent community planning

The Prince's Foundation for the Built Environment (PFBE) – West Winch and North Runcton

The potential for the growth of King's Lynn, specifically the location of up to 6,000 new homes, was being explored by King's Lynn and West Norfolk Borough Council as part of a wider strategic planning exercise.

PFBE facilitated a one-day workshop to examine issues and opportunities in the context of the wider growth agenda. Residents of villages to the southeast of King's Lynn, where expansion was most likely to occur, along with parish, borough and county councillors and officers, were invited to a workshop called 'Exploring Sustainable Growth', which was held at a local social club.

The workshop began with introductions from key stakeholders and presentations on technical issues. This was followed by a bus and walking tour which allowed participants to see the area and consider the scale, location, constraints and opportunities.

In the afternoon, PFBE introduced principles of sustainable growth through past projects and exemplars. The attendees were divided into groups and 'scoping' workshop discussions (part of the Foundation's Enquiry by Design process) then explored the issues and opportunities relating to social, natural, economic and built aspects of the area – themes which together formed a clear and holistic overview of what must be addressed in finding a way forward for growth.

Groups reconvened at the end of the day and, as a means of clarifying and prioritising the issues raised, the workshop was concluded with participants indicating priorities from all the issues raised using the Concerto model. Concerto, developed by Woolley Project Management Ltd, is a consultation tool which captures all the views of the participants in a fully auditable, measurable and transparent process. It combines a preliminary impact assessment form and a stakeholder questionnaire and uses a computer programme to analyse the collected data. This can be interrogated in a number of ways. In particular, it can show in graphic form the ranking of all issues and highlight those proved to be most important. The Concerto method complements the Enquiry by Design methodology, as the more individual and reflective questionnaire answers can be compared (usually, as in this case, very favourably) with the results from the group mapping and discussion sessions, but in a far more focused manner.

Overall, feedback emphasised the need to integrate the old and new development in harmony, to reduce through traffic and ensure careful master planning to create a more 'pedestrian-friendly' village. The feedback produced a valuable evidence base for the future master planning for the area, in which many members of the community are now inclined to participate. The information was used to inform King's Lynn and West Norfolk Borough Council's core strategy, and the suggestions made were largely incorporated into the local plan.

The process of participatory planning can vary hugely, but a typical approach is as follows:

1   Pre-engagement research and dialogue
2   Training for community groups/stakeholders in planning and facilitation as required
3   Public exhibition or meeting
4   Community planning day or weekend of 'hands-on' planning at which groups of residents, assisted but not directed by professionals, create visions and solutions, which they then feed back to the larger group
5   Meetings between the development team and various local groups, or a dedicated stakeholder group, to prioritise ideas and confirm vision
6   Master plan developed by professionals following local insight
7   Further consultation using exhibitions, meetings or perhaps further workshops
8   Final master plan/report produced

The benefits of this approach are substantial. Early engagement engenders community spirit, builds trust between the development team and local community and generally results in positive sentiment towards change. The process creates a sense of ownership among the community, which has been seen to bring about very positive community relations over the long term. The resulting master plan benefits from local insight and, when the process is well coordinated, can combine local passion with professional expertise to produce innovative ideas. Participatory planning, because of the variety of tactics and the emphasis on facilitation, can involve a wide range of local voices, including those who may not choose to comment on a more formal consultation. The process has also been shown to accelerate master planning because it involves not only local residents, but also politicians and planners. That said, a considerable amount of time must be invested at the outset for these techniques to succeed.

Participatory planning is well suited to projects at an early stage in the planning process where there is an opportunity for the community to be involved in developing a vision. It has flourished with the advent of neighbourhood planning, and large-scale mixed-use schemes increasingly benefit from this approach.

The process works less well for smaller schemes where there is little opportunity for flexibility or a short time frame. Several professionals are needed to provide a hands-on role, because considerable expertise is needed to balance the views of local people – particularly in encouraging the less articulate to have a voice and prevent the project from being dominated by certain individuals.

Local authority officers are expected to invest time in developer-led consultations and in attending workshops, providing local insight and resources and assessing the outcomes in relation to wider local issues.

Box 16.2 Case study: the use of charrettes

Marshalls (Holdings) Limited – Wing, Cambridge

Wing is a proposed development of 160 acres on land east of Cambridge and owned by Marshalls. The master plan envisages up to 1,300 new homes, public open space, including sports pitches, tennis courts and allotments, a new primary school, nursery and community facilities, a food store, replacement petrol filling station and other local shops. At the very start of the project, it was decided that the development should reflect landowners' core values and be something that the Marshall family would ultimately be proud to have as their 'back garden'.

Enquiry by Design is not widely used around Cambridge but is embraced wholeheartedly. Over 1,000 people including vicars, parish councillors, doctors, the Cambridge Cycle Campaign and even estate agents took part over three days in 12 sessions, layering thoughts from desire lines across the site to community facilities, parks, play spaces and finally roads and homes. The development team then invited councillors, officers, staff and members of the public to come and comment on the initial concepts.

These were refined, and the development team continued to consult with the community. The local parish council and council members were engaged throughout the process.

The development team also set up a public art group and integrated the products of its work into the master plan. After a year, Marshalls

submitted the planning application based on a master plan which had
been designed by the public for the public.

Fewer than 15 people objected to the application, and Marshalls
has attributed this to the care that was taken to involve the local community and the extent to which they were prepared to engage positively with the proposed development.

Commenting on the process, property director Emma Fletcher said,

> Yes, it was more expensive upfront to undertake an Enquiry by
> Design but it brought other unforeseen benefits: the team and
> community bonded quickly because we were all "in it together" for
> the three initial intensive days, there was a greater understanding
> of the local area from the very beginning, and we gained a net-
> work of people who we could quickly bounce ideas off while the
> design process was underway. We felt that the public had some of
> the best ideas because of their complete understanding of their
> neighbourhood. Possibly it's not suitable for every scheme and
> yes, it takes a certain leap of faith, but would I do it again . . . on
> every scheme I can.

Online consultation

The increased popularity, power and availability of the internet accounts
for much of the change in public consultation over the past two decades.
This increased significantly in 2004 when Web 2.0 enabled more effective
two-way communication. The rise in individual and community use of the
internet, combined with the requirement in 2005 that local authorities
and other public bodies 'e-enable' all services including planning, public
involvement and consultation, means that all development proposals have
an online presence.

There are many other reasons why developers increasingly choose to use
online consultation:

- **Research:** The internet is by far the most powerful research resource.
  A substantial proportion of information that is required in researching stakeholder groups and necessary background information is
  freely and readily available. A wide range of publics can be identified
  online – initially. The internet may supply up to 90% of the stakeholder information required for a consultation, but the remaining
  (and very important) element is often best addressed through personal contact.

- **Issues management:** A constructive consultation is based on the community having access to reliable information, which can be easily sourced online. Monitoring of online consultation provides an immediate and effective means of understanding local sentiment and identifying any misapprehensions.

- **Immediacy:** Online consultation has the advantage of being immediate – information can be posted and responded to in minutes. But consultation timelines should not be shortened as a result. On the contrary, immediate communication can only take place if the audience has been targeted and is in receipt of the message. Online communication can potentially spread quickly but only if the message is strong and compelling.

- **Ease of access:** Online communication is a medium in which many people choose to communicate and by targeting residents via their preferred means, the likelihood of involvement is increased. Users can take part in an online consultation when and where they want – at home, on the move, while waiting for something/someone. Many choose to take part in consultations late at night. Because of its increased accessibility, online consultation has the power to reach new audiences – particularly the young and the time-poor. Local authorities welcome developers' inclination to consult more widely; simultaneously, this enables developers to unearth the support of the 'silent majority'.

- **Dialogue:** Online consultation allows for real-time dialogue and an exchange of ideas on a one-to-one, one-to-many and many-to-many basis.

- **Removing hierarchies:** Online consultation is capable of removing hierarchies. In a busy public meeting, for example, attendees may defer to a dominating character or group leader. Ultimately, those members are not adequately represented, despite their presence. Online, and particularly behind the veil of a username, individuals are more likely to voice opinions without fear of repercussions, while personal details remain confidential but are accessible to the local authority as a confidential appendix to the consultation report.

- **Reaching hard-to-reach groups:** Another reason for the sharp increase in online consultation is that many people – particularly commuters, families with young children, the elderly and disabled – are not easily able to attend consultation events. Online consultation provides an alternative, accessible means of engagement. Online consultation is accessible in both its language and in the varied ways in which information is presented. Voice recognition is breaking down barriers and enabling people to communicate online in the way which suits them best. 'Translations' of complex technical documents as well as translations into different languages can be made available, and the inclusion of email addresses and phone numbers enables users to obtain clarification should they require it.

- **Promotion:** Social media enables messages to be communicated quickly. Blogs and the local media are also integral to most consultations.
- **Feedback:** A consultation website, email and social media provide ideal means for communicating feedback.
- **Moderation:** The way in which a consultation is to be moderated should be determined at the start and communicated via a user guide to ensure consistency. The consultor will need to determine whether user-generated content is to be vetted before appearing and, if so, on what basis comment might be withdrawn.
- **Analysis:** Online communication can be very effectively analysed, comprising day-by-day website usage; average session times and bounce rates; analysis of the most popular pages; demographic information in relation to location, gender, age and interest; analysis of how people are reaching the website; results per poll/forum/survey/blog comment; and maps to depict the location of respondents. Likewise, qualitative analysis which combines a technical and human approach can be more sophisticated than offline analysis.

Taking this into account, it is not surprising that development teams, rather than considering *whether* to run an online consultation, should focus on *how*.

Box 16.3 Case study: online consultation

Scotch Corner Designer Village – Polity and ConsultOnline

Scotch Corner Designer Village is a proposed designer outlet village to be situated at a key junction on the A1(M) in Richmondshire, North Yorkshire. An extensive consultation was run by public affairs advisers Polity, using ConsultOnline for the substantial online elements of the consultation. Following a positive local response, the scheme gained resolution to grant planning consent in January 2015.

Offline community involvement events included an exhibition held on three dates at venues throughout the area (residents were invited by letter), a comment form at the exhibition, a private briefing session for councillors and considerable use of local print and broadcast media, which not only communicated key messages but encouraged discussion.

Online consultation was deemed a necessary part of the consultation because of its ability to reach a wide geographic and demographic profile quickly and efficiently, to target the young as well as the old, to convey information through a variety of means (visual, verbal, audio, video), to bring about positive dialogue, to enable people to communicate at a time and place of their choice and to produce reliable data.

ConsultOnline had a period of six weeks in which to complete the online consultation, working closely with Polity. The scheme's website, www.scotchcornerdesignervillage.com, included questionnaires, polls, forums, blogs and the option for users to post images. Information was available as text, video, images, documentation, email updates, Vcards, Google maps and a blog. The service was monitored 24/7 – enabling ConsultOnline to become aware of, to understand, and to correct misconceptions immediately and for those taking part to receive a quick response. A user guide provided step-by-step advice in addition to terms and conditions.

The website was immediately promoted through Facebook and Twitter: ConsultOnline posted on a wide variety of local organisations' Facebook pages and blogs and established useful contacts via Twitter.

On its first day, the website attracted over 700 unique users and, over the next six weeks, over 3,500 hits.

ConsultOnline provided comprehensive analysis and reporting using Google Analytics, WordPress data and other web-based tools including heat mapping, which featured in the consultation report.

Over two-thirds (77.88%) of users accessed the website using mobile and tablets and 27% of hits were returning users. Almost half (49%) of users found the website via social media; 36% accessed it directly and 13.5% through an organic search.

Demographics revealed that 33.5% of the users were in the 25–34 age bracket, followed by 27.5% aged 18–24, 15.5% aged 35–44 and 11% aged 55+. Men were slightly more likely to take part in the consultation than women (54.15% and 45.85% respectively).

A poll on the website's home page revealed that 92% of those who registered supported the development of Scotch Corner Designer Village.

Comments received both by email and on the FAQ section of the website focused on a range of topics, including the traffic impact assessment, the nature of the shops provided, the possibility of renting office space, sustainability initiatives and parking for disabled people.

Comment was also generated through forums on issues including preferred café/restaurant chains, the popularity of a crèche, preferred designers and whether the scheme should include a play area.

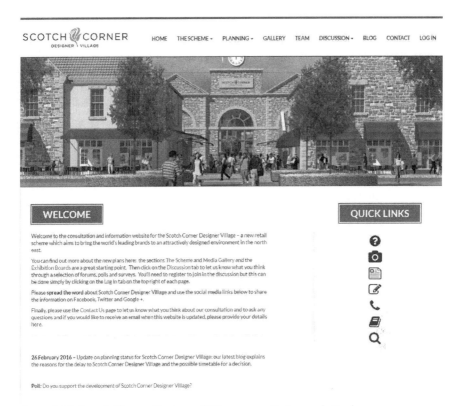

Image 16.1 Scotch Corner Designer Village consultation website: home page

An important consideration is whether to run a consultation solely online, seek a mixture of online and offline tactics or run the consultation entirely offline. In reality, most consultations (and indeed most tactics) now involve an element of online and offline: it would be inadvisable to host an offline event such as a public exhibition without announcing it online; likewise, a website or social media campaign would benefit from being promoted both online and offline.

Smartphones are increasingly popular in consultation, particularly with the time-poor who can use their phone or tablet to take part in an online consultation anytime, anywhere. Not only are most websites now fully responsive (meaning that they adapt to suit all devices), but specifically designed apps enable consultation via smartphones and tablets.

Although it does much to benefit consultation, online consultation is not a panacea: this new selection of tactics presents a new set of risks. The fast dissemination of information online, although beneficial in many circumstances, can also be a disadvantage. In cyberspace, information can fragment quickly and become used by pressure groups to reinforce their interests and prejudices: the energy sector (particularly in relation to nuclear power and fracking) and transport planners have experienced extreme action by protestors exacerbated by the internet. The internet should be consistently monitored and procedures put in place to respond swiftly when necessary.

Furthermore, online consultation, particularly social media, can be seen as superficial and lacking in the emotional power and empathy that face-to-face communication can bring. Online profiles can mask identities, and if measures are not put in place, it can become impossible to monitor the geographical origin of comments. Standardised response mechanisms give online consultations a bad reputation and should be avoided in most circumstances. And despite the increase in online communication, a digital divide still exists, particularly affecting black and minority ethnic (BME) and older groups.

Certainly, online consultation should not be used as a means to reduce costs or labour. As with any tactic, the decision to use online consultation should relate directly to the consultation's overall objectives, including the need to produce meaningful analysis.

Box 16.4 Case study: use of digital technology in consultation

Land Use Consultants (LUC)/REG Windpower – Knightly Hall, Staffordshire

In March 2014, LUC ran a consultation on behalf REG Windpower which included three public information days in Woodseaves, Norbury and Gnosall, Staffordshire, and used digital technology to help to communicate the potential landscape and visual impacts of the development to local residents.

Attendees could view a computer-generated three-dimensional (3D) digital model of the development site and surrounding area. The model was created using detailed topographical data (Ordnance Survey Terrain 5) with a map drape (Ordnance Survey Vector map 1:10,000) and included the proposed wind turbines.

The model allowed interactive exploration: members of the public could identify specific locations on the map and view the 3D model from these locations. Often they wished to look at views from their own property, a particular local vista or a walking route which they valued.

Model views were most commonly set to 1.5 metres above ground level to closely replicate an eye-level view, but equally the height value could be modified to demonstrate the likely visibility from an upstairs bedroom window. The model, which used a bare earth terrain and did not include buildings or trees, provided viewers with a maximum case scenario relating to the visibility of the turbines.

On an adjacent computer monitor, Google Earth was used in parallel to the 3D digital model. This allowed the proposed wind farm, which had been imported as a 3D KMZ file into the Google Earth software, to be viewed from the same selected locations, and in many cases the views could also be switched to a photographic streetview. The photographic streetview provided a more familiar context and the additional level of detail such as trees, hedgerows and buildings appeared to aid the interpretation of the view. Turbines shown to be visible in the bare earth digital model were often found to be screened by tree cover or intervening buildings in Google Earth, thereby lessening the viewers' concerns and providing a degree of reassurance.

How can be online consultation be made to work affectively?

Plan

- Have a content plan in place – but be flexible.
- Watch and listen – determine what works best for the particular consultation, when to post and lengths of posts. Google Analytics is a very helpful tool for understanding user patterns.

Research

- Use stakeholder research and analysis to gain an understanding of the likely take-up.

Use a consultation mandate to establish aims and objectives and guidance on usage

- Ensure that the consultation mandate is displayed prominently – or that its content is expressed clearly.
- Put rules for engagement in place via a user guide.
- Communicate the purpose and process of the consultation. Make the timeline clear and adhere to it where possible; where this is not possible, ensure that the audience is fully informed.
- Be realistic about how quickly you can respond to questions raised online and communicate your commitment to respond at the start of the consultation.

Make access a priority

- Avoid making the online consultation too complicated: always consider the less digitally aware when drafting web content and functionality.

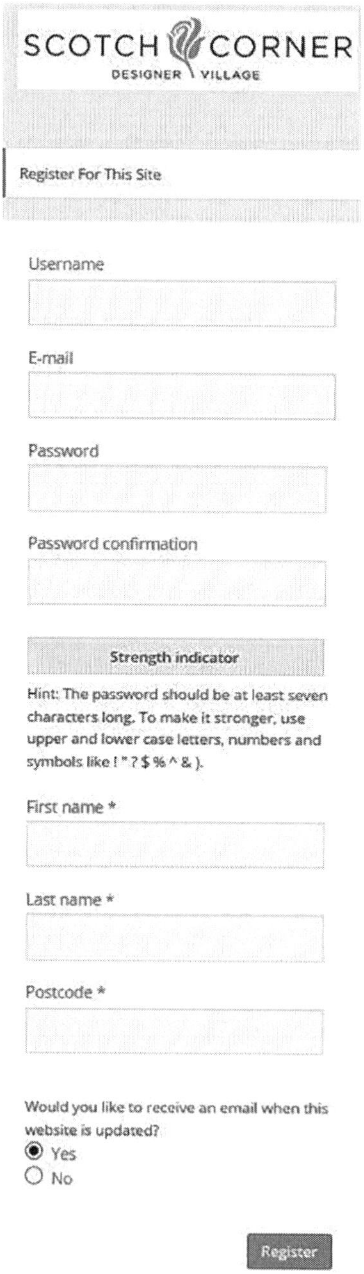

Image 16.2 Scotch Corner Designer Village: registration page

- Consider the benefits of making all (or specific) polls and forums available only to local residents by requiring that they register using a postal address. The importance of registration is three-fold:

  - The proposed development will have a greater impact on those in a specific local area, and so it is important that local residents are given a priority in shaping the proposals.
  - The more detailed the information from the local community, the more value it has. If a developer understands not only what the community feels, but where certain views originate geographically, results are more valid.
  - The consultation report will have added validity if responses can be identified by individual and location.

- Bear in mind that registration can deter involvement. If using a registration process, ensure that this is quick and simple, and doesn't demand so much information as to be off-putting.
- Let people register and get started quickly. Only those with a strong objection to a proposal will persist with an onerous registration process.

Select tactics with careful consideration

- Use a variety of online tactics, providing the tactics are in line with the consultation objectives and deliver meaningful results.
- Aim to use a combination of qualitative and quantitative tactics online.
- Ensure that all tactics, where possible, include an opportunity to respond – if only linking to a Contact Us page.

Create compelling and useful content

- Create an enticing home page. Consider the use of video as an icebreaker.
- Bear mind that people have shorter attention spans online. Write content specifically for the website: do not be tempted to simply install the content of a document or leaflet online (although there is no harm in including these documents in a document library).
- Ensure that text is crisp and clear at all times.
- Break substantial information into manageable chunks.
- Ensure that information is presented in a variety of different ways.
- Provide enough information to enable people to make an informed response.
- Create content that is suitably compelling for people to engage with and share.
- Use images, illustrations, maps, videos and slideshows to bring the content to life.

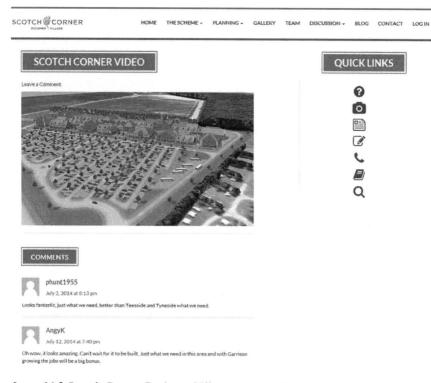

Image 16.3 Scotch Corner Designer Village: project video

- Link surveys and forums to background information to ensure that those responding are adequately informed.
- Provide ample visual material. Mapping can enable residents to zoom in on an areas in detail and add text, video and comment.
- Consider the use of slider bars. This is a visual and effective means of determining relative levels. It works well in budget setting but could also be an engaging and useful tool for community input in landscape design or other decisions.

Blog: a powerful way to provide regular updates and invite responses

- Post regularly and on behalf of various members of the development team, but determine how comment on blogs will be fed into the analysis prior to permitting comment.
- Consider allowing members of the community such as representatives of a stakeholder engagement group to blog.
- Ensure that those who blog on behalf of the development team understand the key messages and the scope of the consultation.

Use information to demonstrate transparency

- Document libraries can be used to hold complex planning documents such as relevant local planning policies, or at the end of the process, the documents which make up the planning application.

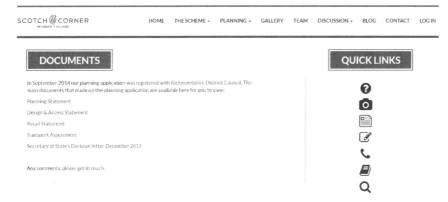

Image 16.4 Scotch Corner Designer Village: document library

- Use hyperlinks to enable consultees to access extensive information if they choose to do so (ensuring that the hyperlink opens a new window rather than taking the residents' attention away from the consultation website).

Involve via online forums

- Use online forums to invite comment and discussion on a range of issues.
- Prior to launching an online consultation, determine to what extent the development team will interact and if so, whether to do so in a corporate character or an individual's name.
- If taking part in online forums aim to facilitate, but avoid arguing at all costs.

Use issues ranking to gain statistical results

- Put in place a mechanism whereby residents can select a preferred option from a list of choices, and second and subsequent lists are selected by routing software in relation to the initial choice.

Use social media with care

- Don't be tempted to use social media just because it's there: consider its original purpose and whether the specific tactic meets the aims and

FORUMS

What's your preferred café or restaurant?

July 31, 2014 1:44 pm 8 Comments

What best revives you during a day's shopping? Would you favour independent providers at Scotch Corner Designer Village? Or is there a chain of coffee shops / restaurants that you'd particularly like to see here? Please let us know!

Would you like to see a crèche included in the proposals for Scotch Corner Designer Village?

July 21, 2014 1:40 pm 3 Comments

Many popular retail developments include a crèche so that parents can enjoy shopping while their young children are well looked after. Would you like to see a crèche included in the scheme? Please tell us what you'd like to see there. What would keep your children happy while you shop? Which age range should be […]

Which designers would you like to see at Scotch Corner Designer Village?

July 2, 2014 11:09 am 11 Comments

Please use this forum to tell us a bit more about what you'd like to shop for at Scotch Corner Designer Village. Which designers would you like to see based here? Is your priority fashion, gifts or home furnishing?

QUICK LINKS

Image 16.5 Scotch Corner Designer Village: forums landing page

SCOTCH CORNER
DESIGNER VILLAGE

HOME   THE SCHEME ▾   PLANNING ▾   GALLERY   TEAM   DISCUSSION ▾   BLOG   CONTACT   LOG OUT

WOULD YOU LIKE TO SEE A CRÈCHE INCLUDED IN THE PROPOSALS FOR SCOTCH CORNER DESIGNER VILLAGE?

QUICK LINKS

Leave a Comment

Many popular retail developments include a crèche so that parents can enjoy shopping while their young children are well looked after. Would you like to see a crèche included in the scheme? Please tell us what you'd like to see there. What would keep your children happy while you shop? Which age range should be catered for? How much would you expect to pay for this kind of facility? If you have had positive experience of crèches on other retail schemes, please let us know the location and what it was that you thought worked particularly well.

COMMENTS

admin@veg2you.com
July 22, 2014 at 10:42 am

Not bothered personally as my children are 5 and 7 however I may have considered it when they were younger. I wouldn't automatically trust someone I don't know to look after my children so a good reputation of trust in this area would take some time to establish.

gac13914
August 18, 2014 at 8:53 am

Personally I wouldn't. I've never left my children at the ikea soft play for example, as I don't know what level of care is provided. An established nursery or childminder has at least been formally ofsted-ed.

AL
September 4, 2014 at 1:13 pm

Maybe a small area like they have at Costa coffee at Teesside Retail Park? Where there are outside benches for you to have a coffee and a climbing frame for kids.

Image 16.6 Scotch Corner Designer Village: online forum

objectives of your consultation. Facebook, for example, was designed as a means to communicate with friends, share photographs and videos and to arrange social activities. It has the means of addressing some consultation objectives, but due to the inability to gain user data and therefore meaningful analysis, Facebook's role as a consultation tactic is limited.

- Likewise, Twitter is a useful means of promoting a consultation, but its 140 character limit restricts meaningful dialogue.
- Bear in mind that many people choose not to use social media, and those that do may not choose to use it to comment on a development proposal. It should not be the sole means of online consultation.
- If you set up a social media profile, keep it active: nothing communicates a reluctance to communicate more effectively than a dormant Twitter feed or Facebook page with unread friend requests and posts. Maintaining a social media presence is a time-consuming process but can be helped by scheduling posts and setting up automatic monitoring, with results directed to a designated email account.

Image 16.7 Scotch Corner Designer Village Facebook page

Image 16.8 Scotch Corner Designer Village Twitter page

Promoting an online consultation

- Bear in mind that consultations are never solely online or offline: successful consultations use online to promote offline tactics, and offline to promote online tactics.
- Use search engine optimisation (SEO) to ensure that the website can be found easily.
- Consider a range of other tactics such as links on other relevant websites, local press and broadcast media, blogs, social media, links on email sign-offs, signage, posters and newsletters.

Be as responsive as is feasible

- Provide a means for respondents to contact a person if necessary – ideally both by email and phone.
- Determine in advance whether you'll interact on public forums – and if so, ensure that the role is one of facilitation, not refereeing.

- Ensure that all those posting/responding on behalf of the consultation do so with the same understanding.
- Respond promptly.
- Keep registered users updated – via email, RSS, SMS or social media.

Remember that communication online is immediate and 24/7

- Commit to regular posting. Social media posts can be scheduled via a range of dashboard applications such as TweetDeck and Hootsuite.
- Keep the website fresh and up to date.
- Check links regularly.
- Update the website regularly.

Ensure consistency throughout the consultation

- Ensure that the online content is in keeping with offline content – this is particularly appropriate if the two parts of the project are being run by different teams.
- Ensure that messages are consistent throughout the website and the wider consultation.

Monitor constantly

- Set up monitoring from Day One. This may be both automatic (for bad language/spam) but should not be exclusively so.
- If you have to remove a post, let the individual know and give them an opportunity to replace it.
- Avoid vetting comments, as this can lead users to question the transparency of the consultation more generally.
- Provide links to offline consultation, allowing respondents to take part both online and offline.

Promote

- Encourage sharing on social networks to help spread the message.

Focus on results

- Avoid the temptation to ask open questions, the results of which may be difficult to monitor and analyse.
- Ensure that the consultation website provides a means of quickly extracting information for reporting and evaluation.

Box 16.5 Case study: virtual reality

VUCITY

VUCITY, the product of communications agency Wagstaffs, GIA and Vertex Modelling, is the first ever fully interactive 3D digital model of London. Through VUCITY, consultees can visualise proposed developments within the existing context of the city. VUCITY can embed transport data, overlay sightlines, identify transport links and sunlight paths and come down to street level to help consultees understand the proposals in context. VUCITY also brings in the London View Management Framework in relation to protected views and can search existing, consented and planned developments.

British Land's Canada Water development is one example of how VUCITY and other interactive 3D digital tools have supported the planning and consultation process: at Canada Water, VUCITY was able to show the architects' proposed master plan as a 3D interactive model in the wider context of central London, allowing consultees to navigate the website and understand the proposals in detail. By showing master plan options, proposed uses, transport links and data overlays, VUCITY was able to demonstrate, in compelling detail, how the proposals would affect the local community.

Image 16.9 VUCITY: an interactive 3D digital model of London

Box 16.6 Case study: social communication tool

Stickyworld/East Architects – Eltham High Street

The increased popularity of social media has resulted in a wide range of social communications tools which may be used for consultation purposes. Stickyworld is a multimedia stakeholder engagement platform that enables the creation of interactive engagement websites called 'rooms', in which organisers present different kinds of media and pose questions, and where participants place virtual sticky notes directly on those media to engage in conversations.

The Royal Borough of Greenwich introduced Stickyworld technology to the regeneration team leading a Transport for London–funded public realm scheme at Eltham High Street.

The project's architects used Stickyworld to complement face-to-face exhibitions and workshops by initiating online conversations, presenting architects' plans and capturing ideas from residents and local businesses. East Architects ran a time-limited online discussion on the opportunities, ambitions and constraints of reshaping the urban realm in the high street. A video from the project director presented the opportunity to get involved, and visitors could choose to comment anonymously or register and join the conversation on a range of issues. In four weeks, almost 400 comments were collected.

Stickyworld was then used to thank participants for submitting their contributions, and PDF and CSV reports were exported and analysed. East Architects then used this and other collected information to inform the development of their ongoing plans. The revised scheme was presented to interested stakeholders in a second 'room', timed to align exactly with the face-to-face consultation. Multiple choice and open text questions were used to mirror the paper-based survey of the consultation exhibition.

The architects believed the use of Stickyworld to be highly beneficial, specifically in reducing the chance of miscommunication and encouraging focused engagement.

Box 16.7 Case study: the use of Minecraft in planning by the University of Dundee–Centre for Environmental Change and Human Resilience/Geddes Institute for Urban Research

Comment by Dr Deepak Gopinath, lecturer in town and regional planning at the University of Dundee, and Lorna Sim, planning officer, TAYplan

TAYplan, the Strategic Planning Authority for the Tay Cities region, was keen to find innovative ways of involving young people in the strategic planning process.

Together with the School of the Environment at University of Dundee and facilitated by the Planning Advisory Service (PAS), we held a youth camp on 11 June 2015 at the University of Dundee. Seven secondary schools across the TAYplan area attended and worked together to develop hypothetical visions for the Dundee waterfront through interactive sessions using the popular game, Minecraft.

Each participating school team was given an individual plot on a model of the Dundee waterfront and encouraged to develop their visions. At the end of the day, we selected a winning team presented them with a Future Planner Award.

A central aim of the youth camp was to encourage young people to think differently about places where they live and to help them understand how they can influence change. We consider the role of young people in planning (both through consultation and in terms of the profession) to be centrally important. Likewise, it is important to help build the future of planning by encouraging young people into the profession. Often, time is not taken to create the opportunities for young people to engage with planning, but planners need to encourage future generations to have more of a say in how their places change.

Minecraft classroom edition[1] is an excellent learning tool that enables pupils to work as a team and engage in a co-production of knowledge in the design of shared spaces. Equally, as individuals, pupils start to make judgements on the quality of spaces they create, i.e. by asking questions such as, 'Is the wall too high?' and 'Are the roads too far off?' and thus are able to make crucial connections between the spatial choices they make and the resultant impact on the well-being of individuals/groups. Thus, these skills emerge as the basic building blocks that professionals employ when creating master plans of our cities.

Conclusion

As communication increasingly moves online, so too will consultation. Ultimately, online consultation can help make consultation fun and potentially address the serious issue of consultation fatigue. Yet online consultation will not replace offline consultation until 100% of any local community is able and willing to communicate online. To some, a screen will never compensate for a human face and for that reason face-to-face contact should not be abandoned.

Note

1 www.education.minecraft.net

17 Analysis, evaluation and feedback

Introduction

Four activities are fundamental to complete a good, transparent consultation but are all too often neglected in practice:

- Analysis of the results of the consultation
- Evaluation of its process
- Feedback in the form of a consultation report (or Statement of Community Involvement)
- Feedback as a courtesy to those who have taken part in the consultation.

The ultimate objective of a team running a pre-planning consultation is to create a set of meaningful results. Analysis is the process by which data becomes meaningful. But it is also important to demonstrate the validity of the consultation itself, which is why evaluation is important. Without demonstrating that the consultation was done well, its results carry little credibility. Finally, providing feedback to those who have taken part in the consultation is an important courtesy.

Care must be taken to ensure that the three roles are not muddled or combined into a single, confusing exercise.

Analysis

Analysis is the collection of data generated by the consultation – percentages from polls, comments from emails, reports from workshops – and the process of making sense of it. This is both simpler and more effective if the analysis of each tactic is planned in advance. A so-called consultation tactic which does not produce data in a form that can be analysed is counterproductive to the consultation: not only is it a waste of resources, but if information is requested which does not then form a meaningful part of the analysis, trust may be destroyed.

Data usually falls into one of two categories: qualitative or quantitative. Quantitative data can be measured by number. Consequently, analysis tends

Box 17.1 Case study: reporting quantitative data

Chelsea Barracks Partnership – Chelsea, London

The Chelsea Barracks Partnership appointed Soundings to carry out a public consultation on the substantial redevelopment of Chelsea

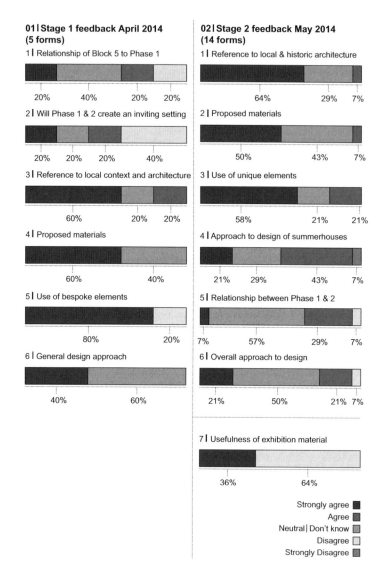

Image 17.1 Extract from the Statement of Community Involvement for Chelsea Barracks

Barracks. Three Statement of Community Involvements (SCIs) were produced which reported on a wide range of consultation activities.

The bar charts in Image 17.1 demonstrate how data collected on feedback sheets from two drop-in events was summarised.

to be relatively simple. Typically, quantitative data may comprise percentages ('67% of those attending the exhibition supported the introduction of a new footbridge'), quantities ('546 individuals supported the proposals') or comparisons ('Five members of the committee voted in favour of Design Option 2, and two opposed it.'). Cross-tabulation can create a more meaningful picture ('Of the 546 individuals who supported the new footbridge, 87% were daily commuters'). Likewise, quantitative data can be useful for comparison purposes ('87% of local residents supported the new footbridge following the announcement of Design Option 2; prior to this, only 61% of residents were favourable') or showing changes in attitudes over time ('Support for a new footbridge has increased in excess of 10% year on year for the past five years').

While quantitative data has the potential to produce clear and definitive results, it can lack depth and works most effectively when combined with qualitative data.

Qualitative data – observations and comments, usually expressed in words rather than in numbers – can provide a context for quantitative data, shed light on the motivations behind individuals' views and enable the consulting body to get to the heart of an issue. But there is no standard form of analysis for qualitative feedback. Not only is it more difficult to measure words than numbers, but it is difficult to do so objectively. Views expressed are invariably contextual, and the context needs to be understood to fully comprehend the comment. Public meetings are notorious for producing extreme amounts of negative comments as a result of local residents being swept along in an anti-development tide by strongly opinionated, and perhaps misinformed, individuals. Where such misapprehensions exist, it is vitally important that they are explained in the context of the comment.

In reality, qualitative responses tend to work best at the early stage of a consultation when issues are being explored, and are less relevant when a local community is presented with a choice of options. Most importantly, the means of analysis should be determined at the earliest possible stage and executed in as much detail as possible, allowing consistency in coding, inputting and analysing the data.

Different methods by which specific tactics can be analysed are suggested in Table 17.1, although it should be noted that these methods are by no means exhaustive. Certainly as far as online consultation (particularly social media) is concerned, new methods of analysis are being developed rapidly.

Box 17.2 Case study: categorising responses to open-ended questions

Argent (King's Cross) Ltd and Fluid Design – King's Cross, London

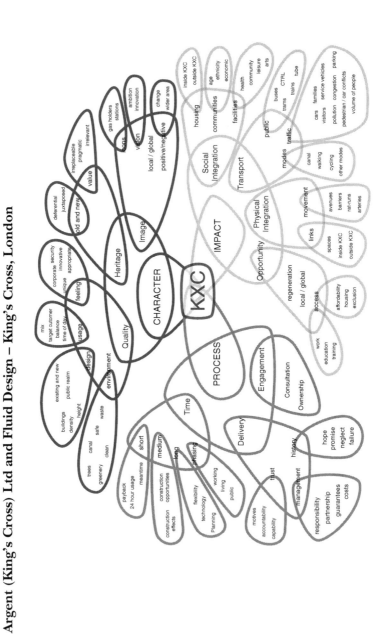

Image 17.2 Visualisation of qualitative research carried out by Fluid Design on behalf of Argent (King's Cross) Ltd.

Table 17.1 Qualitative and quantitative tactics

	Quantitative	Qualitative			
	Percentages in support/ opposition	Record comments and ideas and record based on issue/strength of feeling	Visualisa- tions pro- duced for considera- tion	Group written comments by issue and assess strength of feeling	Record sentiment
Architects in schools		✓	✓		
Art workshops		✓	✓		
Ballots/referenda	✓				
Blog				✓	
Charrette		✓	✓		✓
Citizen's jury	✓	✓			✓
Community forum	✓	✓			✓
Consultation café		✓			
Deliberative poll	✓	✓			
Email				✓	
E-mapping			✓	✓	
Exhibition/ roadshow	✓	✓		✓	✓
Facebook	✓			✓	
Feedback form	✓			✓	
Focus group	✓	✓			
Hotline		✓			
Members briefing	✓	✓			
Models		✓	✓		✓
Neighbourhood forum		✓			✓
One-to-one meeting		✓			
Online forum				✓	
Online picture boards			✓		
Opinion poll	✓				
Participation games		✓	✓		✓
Poll	✓				
Public meeting	✓	✓			✓
Q&A				✓	
Questionnaire/ survey	✓				
Round table discussion		✓			✓
Site visit		✓			✓
Think tank		✓			✓
Twitter	✓			✓	
Video soapbox		✓			
Vox pops	✓	✓			

Box 17.3 Case study: analysing qualitative responses

Transport for London – Crossrail 2

In its consultation on proposals for Crossrail 2, Transport for London (TfL) included an open question which invited respondents to submit any further comments.

In analysing these responses, code frames were developed which were grouped into themes. The code frames enable the number of comments regarding particular issues to be quantified.

As a first step, however, responses were divided up according to the respondents' support of two alterative options, Regional and Metro, as follows:

- Support both Metro and Regional options, or support one and no strong view on the other
- Support Regional and oppose Metro
- Support Metro and oppose Regional
- Oppose both, or oppose one option and no strong view on the other
- No strong view on either option.

The open-question responses were then coded and analysed within each of these groups. The code frames included a series of themes upon which the respondents' comments were grouped:

- Positive comments: unspecific positive comments
- Congestion/overcrowding relief: comments about destinations or areas which would benefit from Crossrail 2
- Improves links: comments about destinations or areas which would benefit from Crossrail 2
- Improves interchange: comments about destinations or areas which would benefit from improved interchange as a result of Crossrail 2
- Reduced travel times: comments about destinations or areas which would benefit from improved journey times as a result of Crossrail 2
- Economics: comments about economic benefits or concerns relating to Crossrail 2
- Environmental: comments about environmental benefits or concerns relating to Crossrail 2
- Social: comments about social benefits resulting from Crossrail 2
- Capacity: comments regarding the implication of rolling stock and route alignment on the services capacity

- Timescales: comments about the timescales and completion of Crossrail 2
- Prefer Regional option: reasons why the Regional option is preferred
- Prefer Metro option: reasons why the Metro option is preferred
- Suggested destinations: comments about destinations or areas respondents would like to be served by Crossrail 2
- Suggestions and service specification: comments about suggestions and changes to the Regional and Metro options
- Would rather improve existing infrastructure: comments about improvements to existing transport infrastructure respondents would like to see, instead of Crossrail 2
- Already has good transport links: comments about destinations or areas along the proposed options which respondents consider as already having good transport links
- Insufficient demand along route: comments about destinations or areas along the route which they consider unnecessary destinations on Crossrail 2
- Policy: policy suggestions
- Negative: comments about problems or concerns the respondents have with Crossrail 2
- Request for more information

To ensure consistency between individual coding responses, the first 50 responses coded by each person were checked. A random check of coding on 5% of responses was also undertaken.

Box 17.4 Case study: taking location into account when analysing results

Brooke Smith Planning (on behalf of Cathelco Group) – Hartington, Derbyshire

Brooke Smith Planning was appointed by Cathelco Group to prepare a planning application for the proposed redevelopment of a former cheese factory site, to provide 39 houses, including six affordable units. The scheme also included the refurbishment of existing buildings on site to provide business accommodation, as well as the provision of a number of recreational facilities for use by the village. The site was in Hartington, a historic village in the valley of the Dove which lies within the Peak District National Park.

Prior to the planning application being submitted, Brooke Smith Planning produced a questionnaire that was circulated to residents of Hartington Village. The questionnaire copies were individually numbered to ensure that only comments from those genuinely living in the area were analysed.

Parallel to this, local residents organised for information on the scheme to be available at prominent tourist locations throughout the village and as such targeted many visitors as well as locals. The information provided by local residents was somewhat misleading and showed the scheme in a particularly poor light.

Brooke Smith Planning analysed and reported back on the questionnaire responses, by way of a Statement of Community Involvement. However, it was also considered necessary to highlight to the council the questionable nature of the information publicised by the local residents and the impact this had on the wider comments received in relation to the scheme.

In particular, Brooke Smith Planning highlighted to the council that a significant number of public responses to the application were received from as far afield as Italy and Switzerland, as well as distant areas of the UK such as Dorset, Ramsgate and County Durham. The relevance of these comments was questioned.

When considering principally the comments received from those living in the village, there was a certain level of support for the scheme and in particular the affordable housing and employment space. However, when the scheme was reported to council members at a planning committee meeting, the perceived strong overall objection to the scheme contributed to members refusing the application. This was contrary to the officers' recommendation that consent should be granted.

Box 17.5 Case study: using maps to show consultation responses

Soundings/Bishopsgate Goodsyard Regeneration Limited – The Goodsyard, London EC2

The consultation on The Goodsyard generated a database of over 1,600 individuals and organisations who have been informed or directly involved in the consultation process.

Soundings used colour-coded maps to show by geographical location both the types of groups (identified by group type) and individuals who had taken part in the consultation (Image 17.3).

The maps were particularly effective in showing that gaps identified during earlier stages of consultation had been targeted successfully.

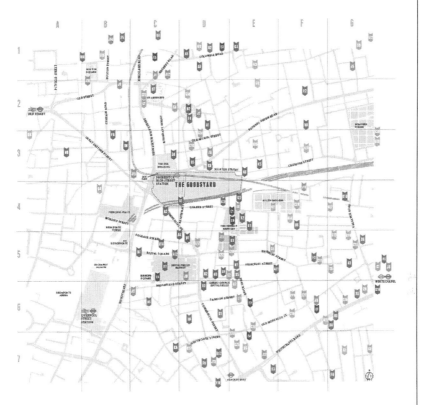

Image 17.3 The Goodsyard: map showing the geographical location of those taking part in the consultation

Box 17.6 Case study: maps showing respondents by postcode

Transport for London – Crossrail 2

In 2015, Transport for London (TfL) ran a consultation on the intention to introduce Crossrail 2 which resulted in 13,767 public responses from across London and the South East.

Maps were used to show support (Image 17.4) and opposition (Image 17.5) for the proposals. A bar chart (Image 17.6) was also used to show responses in relation to specific London boroughs.

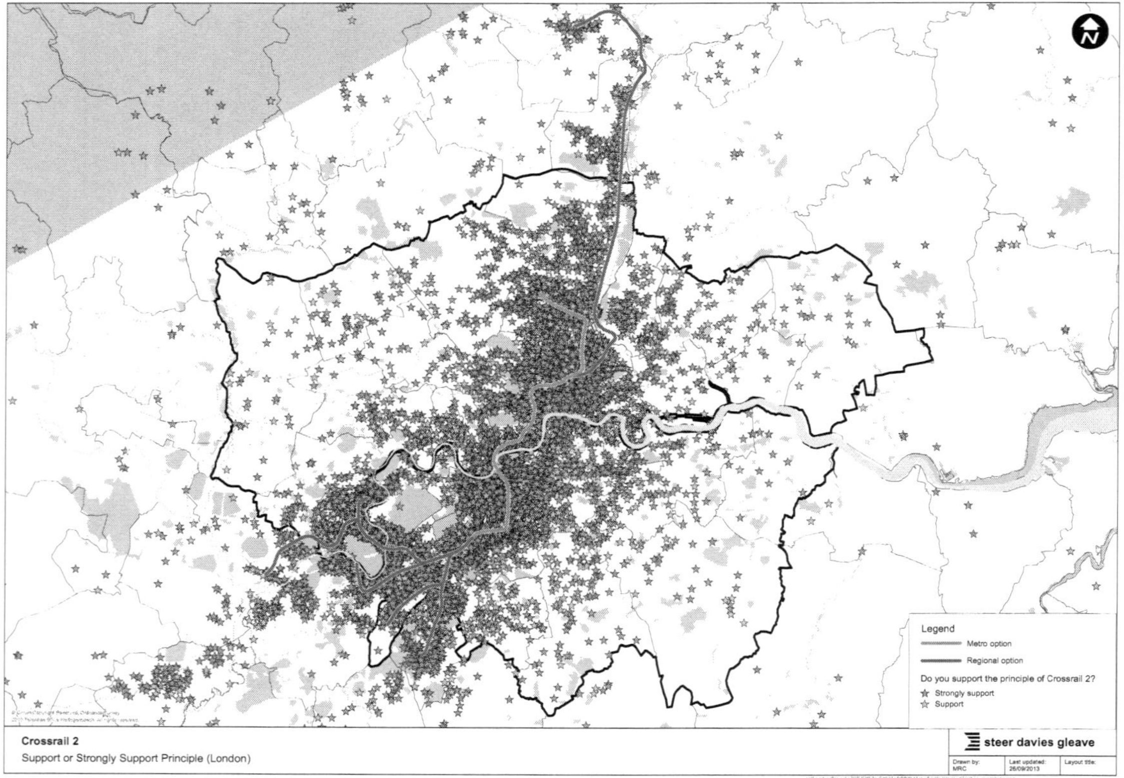

Image 17.4 Crossrail 2: map showing support

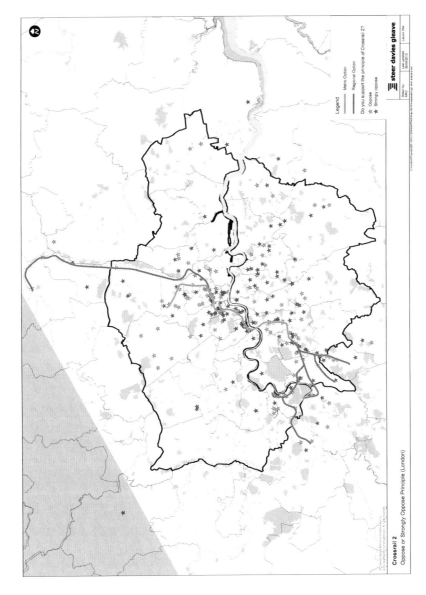

Image 17.5 Crossrail 2: map showing opposition

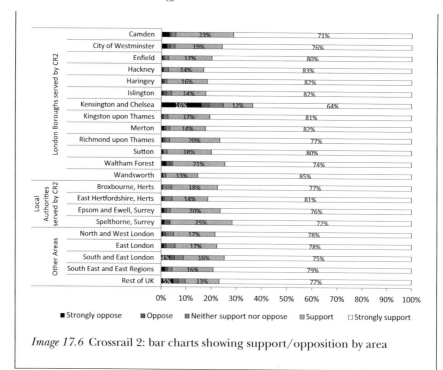

Image 17.6 Crossrail 2: bar charts showing support/opposition by area

Evaluation

Evaluation is the process by which a consultation is reviewed. Its dual purpose is to demonstrate that an effective consultation has been carried out and to benefit future consultations. The former gives credibility to the results and can also make sense of any inconsistencies. For example, initial analysis might reveal that 85% of local residents support the inclusion of an educational facility at a wind farm development but at a small meeting with local residents, only 10% indicated support for the facility. Evaluation of the process would demonstrate that this particular meeting was instigated by the local ramblers group which adamantly opposed any development on the fields in question and thus although accurate, these results were the view of a minority group and, importantly, opposition to the wind farm in principle, rather than to the educational facility.

Although it takes place at the end of the consultation process, evaluation should be considered at the very start. When objectives are drafted, they should be assessed for their potential to be evaluated and dismissed where this is not possible. Chapter 14 covered the setting of objectives and stated the need for objectives to be measurable.

Box 17.7 Case study: an evaluation based on objectives

Queensdale Properties Limited – Hamilton Drive, St John's Wood, London

Objective was to ensure that all consultation materials contain a summary of the plans for Hamilton Drive, a description of the consultation process and contact details

- All materials contained the required information

Objective was to communicate the proposals for Hamilton Drive to all residents within a mile radius

- Residents within one mile were reached

Objective was to engage with 10% of local residents through consultation activities

- Of a database of 842 residents, 126 responded to the consultation

Objective was to gain qualitative and quantitative feedback on design, access and landscaping and adapt the development proposals accordingly

- Substantial feedback was received, with residents whose homes bordered the site engaging positively throughout the development process. 23% of comments concerned parking; 19% concerned architectural style; 10% concerned landscaping; 9% concerned impact of construction work; 7% concerned size and scope of the proposed scheme; 7% concerned support from local residents; 6% concerned potential for light, noise and other disturbances; 4% concerned consultation; 4% concerned provision of affordable housing on site; 2% concerned amenities and practicalities; 2% concerned maintaining the quality of the local area; 2% concerned pedestrian access through the site; 2% concerned questions relating to ownership and financial viability; 2% concerned security issues; 1% concerned environmental issues; and 1% concerned the timetable of work.

Ideally, evaluation is formative rather than summative: making sense of the consultation throughout and making changes as necessary, rather than simply assessing it at the end of the process. There is also an argument for evaluation to be carried out externally to allow for objectivity, although the counter-argument is that the process of evaluation is a useful learning experience for the team at the heart of the project.

Box 17.8 Evaluating the effectiveness of consultation and engagement: Amec Foster Wheeler guidelines

Preparation

- Did everyone (staff, consultees and partners) understand the objectives of the consultation?
- Were the timescales and process transparent and adhered to? If not, why not?
- Were the level and skills of resources right for the consultation undertaken?
- Was the budget adequate? Note areas of overspend/savings for next time.
- Were there any unforeseen costs? What were they?
- How does the cost compare with other similar exercises?
- Was the consultation planned jointly with a partner or neighbouring organisation?

Stakeholder identification

- Were the right stakeholders involved?
- Did you successfully reach all your stakeholders?
- Were you successful in reaching hard-to-reach groups?

Publicity and methods

- Were the publicity materials that you used/developed effective (for example, posters to advertise an event, putting material on the internet, press releases)?
- Did you get the level of information right (for example, was it easy to access, relevant to the consultation, produced in plain language, easy to understand)?
- Was the consultation accessible (for example, were materials available in other languages and formats, where necessary? Were interpreters provided or necessary? Were venues accessible? Did the set-up encourage participation)?
- Were the methods used appropriate for the objectives of the consultation?

- Did some methods work better with particular stakeholders than did others?
- If you used more than one method, which worked best and why?
- Was there the right balance of qualitative and quantitative methods?

Responses

- Were the numbers that took part expected – did you reach your targets?
- Were your response rates high enough to give reliable results?
- Did you get the information you wanted in sufficient time, depth and quality?

Outcomes

- Did the consultation inform a decision or shape policy or an action?
- Did the consultation help set local performance standards and targets?
- Has the consultation helped to improve the cost effectiveness?
- Did it lead to a change of policy or service? (Be specific about how.)
- How many people will be affected by the changes?
- Has the consultation changed the relationship between you, your users and others?

Feedback

- Were results made known to consultees, the wider public, relevant staff and partner organisations?
- Were the consultation findings and outcomes logged onto a consultation database and made available to others who might find them helpful?

Evaluation

- Did the consultation reach a representative sample of the population and all the target groups?
- If the consultation did not meet your objectives, why was this? What steps can be taken to prevent similar problems in the future?
- What was the evaluation of those who took part; did they see the consultation as fair and useful?
- What would you do differently next time?

The consultation report

The consultation report, or SCI, is part of a suite of documents which constitute the planning application. It is often the primary reason for carrying out the consultation, a means of demonstrating local sentiment and a justification for the time and effort invested in the consultation by third-party groups and individuals. While its main purpose is to clarify and demonstrate local sentiment in the case of an outline planning application, it can also explain why elements of the scheme have been retained despite local will.

Analysis should form the basis of the consultation report, with the evaluation illuminating findings as necessary.

Suggested content of a consultation on a large mixed-use scheme is as follows:

1   Introduction

   • Prepared by
   • For
   • Purpose
   • Structure

2   Summary of the planning application
3   Context of the planning application

   • Location – both in terms of local authority administrative areas and roads/other boundaries as a reference
   • Local plans/area action plans/enterprise zones etc. within the locality
   • Size of the site
   • Access to the site
   • Planning history of the site

4   Planning policy context

   • Outline the legislative and planning policy context for undertaking community and stakeholder consultation
   • Make reference to the legislation which specifies the existence of the SCI – the 2011, the National Planning Policy Framework 2012 and the Planning and Compulsory Purchase Act 2004
   • Reference to the local authority's SCI

5   Programme and methodology

   • Outline the approach taken in the pre-application consultation for the proposed development, listing the tactics by both date and category order
   • List stakeholders targeted

- Analysis of those reached (could be broken down into aware/informed/engaged)

6 Outcomes of the consultation process

- Brief descriptions as to how the information was processed and analysed
- Response rates, including demographic information which provides a background to the responses
- Key issues raised in public and stakeholder consultation, by category
- Development response to issues raised

7 Conclusion

- Summary of each section
- The impact of the consultation on the development proposals

8 Appendixes

- Reports from workshops
- Questionnaires together with responses, with quantitative data presented graphically
- Letters and newsletters
- Invitations to events
- Press advertisements
- Flyers and posters
- Consultation statistics – number of hours of public events, numbers of letters issued, numbers of newsletters issued, number of questionnaires received, number of groups met

The following images might be included:

- Maps showing the geographic areas targeted
- Maps showing the location of consultation events
- Maps showing responses in a geographical context
- Press advertisements
- Photographs of the exhibition
- Exhibition boards
- Visual feedback from workshops
- Screenshots from the consultation website and social media pages

The consultation report is a public document. Although the local authority will require information as to who was consulted, it is sometimes necessary to redact personal details, for example, replacing 'Anna Williams, Lannesbury Crescent' with 'Resident, Lannnesbury Crescent'.

Finally, although a developer's consultation will seek to demonstrate support, it must do so through demonstrating that the consultation was genuine, responsive, engaging and accessible. 100% or even 90% in favour of the proposals does not necessarily represent a good consultation; in fact, it may raise suspicions. The consultation report should demonstrate that a positive consultation delivered a positive result: a positive result without the necessary evaluation to determine that the consultation was well managed means very little.

Feedback

The final stage of the process is to update those who have taken part. It is advisable to communicate with consultees in the way in which they choose to communicate – by letter in some cases, by email in others or face to face. In the interests of good ongoing community relations, it is prudent to do so in advance of or at the same time as issuing a press release. Feedback is an opportunity to thank consultees for their time, a means of communicating the results and potentially an opportunity to encourage those who have been supportive to continue in this vein.

Box 17.9 Case study: lessons learnt

Transport for London

Transport for London (TfL) runs several consultations at any one time, ranging from local consultation on changes to bus routes to long-running and wide-ranging engagement programmes, such as the series of consultations on the Silvertown Tunnel.

Following the completion of a consultation, its effectiveness is evaluated and conclusions are drawn. The following subjects are included in the evaluation:

- The suitability of types of venues
- Reasons for non-engagement
- The nature of questions raised in relation to specific consultation events
- Views on consultation materials – content and accessibility
- Means by which people acquired information about the consultation

Additionally, TfL routinely evaluates its consultations in the context of the Gunning Principles, specifically, that consultation should:

- Be held at a formative stage, so that respondents have maximum opportunity to influence decision-making.

- Provide sufficient reasons to allow 'intelligent consideration', so that respondents can come to informed opinions, maximising the value of consultation to them.
- Provide adequate time for consideration and response, to ensure that respondents have sufficient time to come to and express a view, and that there is sufficient time to properly consider that view.
- 'Conscientiously consider' feedback received, so that respondents can be assured that their view has been properly considered.

The evaluation of each project contributes to an overarching document which serves to better inform future consultation strategies.

18 Reducing risk in consultation

Introduction

While the best consultations are those which are transparent and accessible and provide significant opportunities for engagement, these characteristics are not free of risk. The consultor's priority will be to run as genuine a consultation as possible while retaining an awareness of risk and taking steps necessary to mitigate it.

Activism and negativity

Perhaps one of the greatest fears of those running a public consultation is the negativity that can be voiced by local interest groups, national activists and so-called NIMBYs.[1]

Campaign groups are rising in prominence in the UK: 1.5% of the UK population (over 1 million people) are believed to be paid-up environmental group members,[2] and research suggests that campaign groups are more trusted than local MPs, councillors and the local media.[3] Not only are these groups becoming more numerous, but they are also growing in sophistication, knowledge and tactics.

Lobby groups are known to gather supporters nationally to fight a local cause, transport people to attend public meetings and protests and even set up local branches to ensure that they can infiltrate local opinion – as has been seen on many occasions in the debate over a possible third runway at Heathrow Airport.

As their very name suggests, interest groups unite around a single shared cause for which they are committed to fight. Local conservation societies, for example, are created to protect a specific locality in its current form and consequently may resist change regardless of its potential benefit. Conservation societies are typically very well run, have a committed and well-educated membership which may include politicians, and will have campaigning experience and access to legal and other resources.

It is common knowledge, particularly within the property, transport and utilities sectors, that individuals are more likely to rally in objection to a proposal than in support of it. Opposition is perhaps the quickest way to unite a community and as passions run high, awareness of and support for a cause take hold quickly.

As Chapter 5 explained in detail, the internet is raising the stakes further, with individuals and groups able to set up web-based platforms and social media accounts as easily as can developers. Campaign to Protect Rural England (CPRE)'s Planning Help pages exist to enable people to take action on planning applications and include sample letters of objection. Architects and planning consultants who feel their employers' proposed schemes lack social value are encouraged by the organisation Concrete Action to upload plans anonymously as the basis for a local campaign. Similarly, organisations such as Better Transport encourage individuals to campaign against NSIPs in their area.

The phenomenon of NIMBYism has developed in response to the perception that development will threaten established neighbourhoods and is fuelled by the common human characteristics of resistance to change, nostalgia and fear of the unknown. The term is sometimes used to describe all local residents who oppose a planning application. This is misleading: the label NIMBY should, if used at all, only refer to those whose objection is based on selfish motives, is subjective and, because it extends beyond material planning considerations, falls outside the remit of the consultation.

Despite there being some strong external forces in opposition to development, the development process itself is partially to blame for some negative sentiment. A concern commonly voiced by property developers is that local plan consultations frequently fail to consult fully and adequately. Land is then allocated for development unbeknown to local residents and it is only when a planning proposal is put forward that its neighbours are aware and raise an objection. At this point the principle of development has already been established and the developer's consultation meets its first hurdle.

What can be done to minimise the impact of any ill-founded issues that activists, special interest groups and so-called NIMBYs can bring about, and how can these groups be to engaged more positively? Effective use of the strategy can mitigate many of these issues.

Using research

The very first stages of the strategic process (described in Chapter 14) present the first opportunity: research and situational analysis can gain an understanding of local groups (and national groups where relevant) and their motivations and methods. Initial research should help identify possible misapprehensions which can then be addressed as the strategy is developed and implemented.

Box 18.1 Correcting misapprehensions

Comment by Paul Butler, PB Planning

PB Planning has been appointed by a number of clients to consult on proposals to deliver new housing developments in a variety of edge-of-settlement locations: sites which are located within the open countryside, Areas of Outstanding Natural Beauty and the green belt.

PB Planning undertakes initial research which reveals whether the proposals are likely to meet with concern from local residents who may enjoy the land in its current form and are resistant to new homes being built.

Further research is used to demonstrate the need for the development. Messages are developed which include the following statistical information:

1   Approximately 75% of the country is greenfield, of which nearly half is farmed for agriculture. If the government's target of an additional 200,000 homes per year were met, it would reduce greenfield land by a small fraction (circa 1–2%) over a period of five years.

2   The local plan is cited, specifically local housing demand and analysis demonstrating the suitability of the site for housing.

3   Positive messages are created around job creation, again drawing on statistics. Housebuilders have the capability to calculate how many construction and related jobs a specific development is likely to generate. The economic benefit of the scheme can then be expressed in monetary terms, providing both demonstrative and statistical reassurance.

Having met with many local residents, PB Planning also becomes aware of negative comments in relation to the planning system: there is sometimes an assumption that the planning application would automatically gain consent. Again, PB Planning tackles this issue with easily digestible figures, in this case relating to scoring in sport: for developments to be approved and then constructed, the applicant needs to win '10–0', i.e. ensure that the local planning authority and all of the statutory consultees are in approval of the scheme. If just one of these parties are in objection to the development, then the development will not be approved, i.e. '10–1' is a loss to the developer.

Early stakeholder research and PEST and SWOT analyses are quick to identify issues, and addressing misapprehensions early on usually results in a scheme being welcomed by local residents and planning consent granted.

The consultation mandate

The consultation mandate as a means of crystalizing the strategic approach provides an opportunity to convey the principles of the consultation – such as its intention to be responsive, genuine, engaging and accessible. If it achieves this, it can provide some reassurance to potential opposition.

Importantly, the consultation mandate is the best opportunity to put in place the rules which govern the consultation. Should it be necessary, the mandate can state that the consultation team may remove negative comment from discussion boards legitimately and dismiss irrelevant comments. For example, if it is made clear that the consultation forbids bad language or verbal attacks against individuals, the consultation mandate is justification for removing such a comment. The consultation mandate will also describe methods of analysis. Where results are analysed by issue rather than by number, a postcard campaign or automatically generated email will carry little weight. The method of analysis should be clearly stated in the consultation mandate to avoid the escalation of negative sentiment.

The consultation mandate may also stipulate the boundaries of the consultation, in some cases ensuring that debate focuses on specific questions and options and will not take into account comments which do not respond to the consultation brief. This approach should only be used where there is a substantial danger of opponents dominating the consultation, changing its direction and introducing subjects which are not relevant; in some circumstances, this would be seen as too controlling, preventing discussions from developing and ideas being generated.

The consultation remit

Conversely, it can sometimes be beneficial to extend the remit of the consultation, as greater involvement can create greater empowerment. Elsewhere in Europe, many renewable energy proposals are community-led and are flourishing as a result. This can be effective in the case of wavering opposition: where a *potential* opponent is also potentially a supporter given that certain demands or assurances are met. Many critics have converted to ambassadors simply through receipt of useful and reliable information.

Messaging and questioning

The need to communicate key messages is of paramount importance. An early focus on communicating messages provides an opportunity for development proposals and their benefits to be promoted and substantiated with statistics such as local plan allocation details, housing allocations or the relevant national planning policy.

Messages should also make reference to pre-consultation dialogue – such as the fact that the local authority and its members contributed to the consultation strategy. This can potentially address some criticisms of the consultation, certainly in a political context. Additionally, if councillors put forward a negative view, this will be identified in the pre-consultation dialogue, allowing the consultation team to either adapt the proposals or to draft appropriate responses.

The response 'none of the above' is rarely an option in consultation. As we have seen, consultations vary significantly from those which start with a blank canvas and elicit thoughts and ideas from a local community to those which present a draft master plan and a set of options. In either case, the response sought from the local community will be clear from the early stages of the engagement process.

Asking the right questions is a very important means of reducing negativity. In many cases – for example, in NSIP consultations or other situations where the extent of the consultation is limited – questioning must be tightly controlled to avoid discussion focusing on larger and potentially controversial issues. An NSIP project should not seek general views on the proposed development, as it is not within the remit of the consultation to address the *principle* of development. Instead the consultation will focus on very specific questions, such as design options and discussions about community benefits. Where necessary, the consultation mandate should clearly stipulate that it is not within the remit to consider *whether* the development is to go ahead, but *how*. Contentious subjects such as new housing can also be mitigated by adapting or rephrasing questions such as 'Do you think housing should be built on this field?' with 'Do you think that people would like to live here?' Similarly, if identified, difficult issues can be addressed in the questioning, for example, 'If the development was to include a roundabout to alleviate traffic congestion, would you support the proposed plans?'

Monitoring, analysis and evaluation

With the potential for difficult discussions and negative feedback, there is an acute need to monitor and perhaps guide discussion. The extent of the development team's involvement in listening and taking part in discussions should be established at the start of the consultation and agreed across the team to ensure consistency. In some cases it may be necessary, for example in online forums, only to use moderation in the case of irrelevant, obscene or defamatory remarks, while in other cases it might be felt necessary to focus discussions on the agenda and correct misapprehensions. Consultors should never be tempted to enter into debate.

Development teams should bear in mind that it is not only the proposals that are open to scrutiny: an objector with an axe to grind is likely to fault the consultation too. Conversely, even a potential supporter, when frustrated by the process, can develop negative views towards development proposals. Consultation strategies must be assessed to ensure that they stand up to scrutiny and their merits communicated in a consultation mandate. At the end of the process, the evaluation will demonstrate whether any criticisms about the process were justified.

The use of an issues database

As has already been shown, engagement on proposals likely to attract negative comment may need to be more tightly controlled than less contentious consultations. In consultation, there is often little need to respond to points made until the end of the process; however, where contentious issues are raised and misapprehensions spread, it is often necessary to respond to emerging issues in a public forum and thereby address potentially problematic issues without delay.

It can be hard to change feelings with facts, but facts can certainly help. An issues database, whether publicly accessible or kept for reference purposes, can include figures relating to statistics (increased traffic figures in the case of a new road, housing allocations in the case of a residential development) and financial data (employment opportunities and the value of the new scheme in monetary terms).

Box 18.2 Case Study: issues register

Transport for London (TfL) – proposed changes to bus routes in Hounslow, Ealing and Twickenham

Public consultation occurs on each TfL project where changes are proposed. Public consultation enables the local community to influence that change. As part of this process, a 'responses to issues raised' document is produced.

This document publishes the issues raised and lists the TfL response. This enables consultees to find all the answers to issues raised in one place.

Every issue raised is covered and responses are prepared jointly by the project and consultation teams.

Consultees who have supplied contact details are emailed a link to the issues register online. Additionally, hard copies are posted to those who request them.

Although issues raised are generally more likely to be of a negative nature (positive responses are more likely simply to state their approval than to raise issues), the issues register is a very constructive element of the consultation, showing how the project is intended to progress in relation to issues raised and ensuring full transparency.

Selecting tactics

Selecting appropriate tactics can help create the right environment for positive discussion: tactics should be arrived at, their suitability having been discussed in pre-consultation dialogue to prevent criticism of the consultation itself; tactics should be sufficiently engaging to encourage the 'silent majority' to take part and information tactics should be used to reduce the potential for the message to spiral out of control. Unfortunately, activists have been known to create misapprehensions deliberately to introduce fear and confusion; ensuring that accurate and positive messages are always accessible is the best means of combatting this.

Use of a 'middle man'

Difficult consultations have been known to benefit from the involvement of an objective individual who can help the community navigate the options before them without being seen to owe allegiance to the development teams.

Box 18.3 Case study: the appointment of a planning consultant to benefit local understanding

Peter Brett Associates – Alconbury Weald, Cambridgeshire

Peter Brett Associates (PBA) worked with Urban and Civic to develop an Enterprise Campus, 5,000 homes, 700 acres of green open space and a range of transport, energy and community facilities on a former airfield.

During PBA's initial involvement, both the planning consultants and the engagement team at Urban and Civic developed constructive relationships with local people. In the case of the local parish council, however, it was felt that a lack of understanding of the planning process limited effective engagement.

The decision was taken to appoint and fund an independent planning consultant to work directly with the parish council. Working on a part-time basis over a two-year period, the planning consultant, who was local to Cambridgeshire, attended parish council meetings and worked with members between meetings to explain complexities and nuances. This enabled the parish council to gain a deeper understanding of the proposed changes. Furthermore, the parish council's trust of the development team increased and dialogue with the development team was significantly more effective.

Negotiation

Negotiation is a common feature of consultation, albeit in varying degrees. Again, the consultation mandate should recognise in which situations negotiation is possible. In a planning application for a specific scheme, viability determinants, such as number of units, and political determinants, such as proportions of social housing, are unlikely to be a subject for consultation, along with restrictions on the land use brought about by either a legal covenant or the local plan. This too should be clarified in the consultation mandate. But there are many subjects which may be open to negotiation and ultimately this helps create a positive environment for future discussion.

Box 18.4 Case study: forging links with potential supporters

Sainsbury's Local – Bruntsfield Place, Edinburgh

Sainsbury's acquired the former Peckham's deli shop on Bruntsfield Place in Edinburgh's Bruntsfield district, a lively shopping area with a strong mix of independent stores, national chains and coffee shops.

Sainsbury's intention to open a convenience shop in this sensitive location became a very hot topic within the local community, with a number of neighbouring retailers concerned about the impact the opening of this store might have on their businesses.

The 'Say no to Sainsbury's Bruntsfield' Facebook page was established and posters objecting to the plans put up in neighbouring businesses. A debate focused around proposals was even aired on a national radio programme.

Sainsbury's engaged with the community and local press. A local traders' and residents' meeting was held and representatives from Sainsbury's attended to present the proposals and answer questions.

The residents' and traders' meeting was well attended and a great number of views were expressed. Sainsbury's offered to work with the local traders to help re-establish the traders association and assist with any schemes aimed at further enhancing the retail offer in the area. Sainsbury's agreed to offer employment opportunities to the colleagues previously employed by Peckham's in the new Sainsbury's convenience store, and this ensured that great customer service was able to continue.

The site was a very sensitive location, and Sainsbury's worked closely with The City of Edinburgh Council to ensure that the store's muted design was in keeping with neighbouring properties and the local area. As a listed building in a conservation area, a great deal of attention had to go to the design of the premises, ensuring that it complimented the character of the area.

When the Sainsbury's Local opened, six colleagues who had previously been employed by Peckham's took up the opportunity to work in the shop. Links with the local community were further strengthened when Sainsbury's chose Radio Lollipop as its inaugural Local Charity of the Year. At the opening of the store, a Sainsbury's employee appeared live on Radio Lollipop.

Local bars have started buying produce from the store and residents have welcomed both the improvement to the high street and the increased competition.

Impact on reputation

Consultation teams should not be overly concerned about the odd negative comment, particularly those which are based on misunderstanding and can be explained in the consultation report. Decision-making bodies understand that negative sentiment is far more likely to reach their desks than positive sentiment. An important consideration as far as the applicant is concerned is the impact on reputation: the fact that a single negative statement, albeit inaccurate, can fuel fear and mistrust more widely. Monitoring, and responding where necessary with credible information and data, is vital to prevent rumour spreading. Meeting with groups and individuals is often the best course of action, and should groups refuse to meet, the consultation report should make this clear.

Box 18.5 A local authority's view on letters of objection

Comment by Rebecca Saunt, planning manager East Cambridgeshire District Council

In the majority of cases, we don't receive letters of support for applications. The majority of the comments received tend to be objecting to an application. However, even if objections are received for an application, this does not mean that it will necessarily be refused.

Each application is assessed on its own merits and the comments/objections received are reviewed by the case officer to ascertain if they are material planning considerations.

If they are, the objections are assessed and addressed during the course of the application. Even if an objection is a material planning consideration, the weight given to it forms part of the assessment of the application. For example, an objection may be received in relation to overlooking, but the distance between rear intervisible windows could be in excess of 30 metres. Our Design Guide SPD states there should be a minimum distance of 20 metres and, therefore, even though an objection has been raised, the proposal would meet the requirements of our SPD and the case officer would be satisfied that there would be no impact on residential amenity and a refusal would not be issued.

A further example is an objection raised in relation to highway safety. If during the consultation process the Highways Authority has raised no objections on these grounds, while the objections would be noted and assessed as part of an application, a refusal would not be issued. If an application receives more negative comments than positive comments, this does not automatically mean that the application will be refused, as the proposal will be assessed against policy and any other relevant guidance.

Petitions can cause considerable harm to reputation, particularly if it becomes difficult to communicate directly with all signatories – perhaps because they are anonymous or come from outside the area. But fortunately local authorities acknowledge that consultation is not a vote and that decisions should not be based solely on numbers. Previous emphasis on the statistical analysis of responses has changed with the advent of various forms of participatory planning. The feedback from planning workshops at which individuals have invested considerable hours, if not days, is of greater significance than some unidentifiable signatures on a scrap of paper.

Analysis and reporting

In most cases, it is in the consultation team's power to decide how to process comment. At the very start of the consultation process, it should be decided whether to collate responses per person or per issue. The latter is used most frequently because it creates a clear picture of the issues arising from the consultation. And although numbers should be attributed to comments made, analysis by issue ensures that a comment made repeatedly will only be listed once, thus giving a petition or automatically generated email as much prominence as a single individual response. Take the example of an online forum in which local residents are invited to put forward views about public art and design at a new shopping centre. The discussion is dominated by members of an anti-development pressure group who join the discussion stating that the developer is driven solely by profit and has no genuine interest in the local community. This single viewpoint would carry little weight in the final analysis: the issue is based on misinformation and is potentially defamatory (and therefore could be removed for legal reasons), it does not respond to either the specific question or the remit of the consultation, and those responding may not be local residents. The only meaningful information that this produces for the purposes of the consultation report, and therefore the planning application, is that landscaping and public art at the new development are not of particular concern to local residents.

As this example shows, negative views are often either individually motivated or promoted by an external group with a specific cause. Frequently, their views do not represent those of the wider community and this must be reflected in the reporting. Similarly, the consultation report need only include material planning considerations. 'It will spoil my view', 'We don't want more affordable housing' and 'The developer is just in it to make a profit' are not material planning considerations.

The consultation report should also take into account representation. If only 20 people respond to a consultation and are 100% opposed to the proposals, is this a negative outcome? If 500 people were contacted and given ample opportunity to respond, it may be concluded that the proposal is of little concern to the majority of the community. The consultation report should consider the view in the context of the wider community and determine whether the low response was due to lack of awareness or lack of concern. If the applicant can demonstrate that all residents were given substantial opportunity to comment, it could be deduced that those who chose not to respond did so because they were accepting of the proposals.

Analysis can be carried out in such a way that opinion from national pressure groups is recognised as such and priority is given to views from local residents. Thus, the most harm caused by national groups is a negative impact on the developer's reputation, which may lead to negative comments from within the community.

Box 18.6 Case study: dealing with negative sentiment

Brooke Smith Planning (on behalf of Central England Co-operative Limited) – Wirksworth, Derbyshire

Brooke Smith Planning (BSP) was appointed to prepare a planning application for the redevelopment of an existing convenience retail outlet and petrol filling station in the Derbyshire market town of Wirksworth.

Initial research into stakeholder sentiment suggested that local residents would oppose a national chain occupying a retail unit in a location which was largely dominated by local traders.

Having communicated the purpose of the development effectively, correcting misapprehensions and framing the development proposal in a planning context, BSP unearthed substantial levels of support for the proposal.

Two public exhibitions were held. Representatives from BSP, Latham Architects and the Co-operative were in attendance to discuss proposals, concept and design and answer questions posed by the public and feedback sheets were available. Comments were collected in a suggestion box so that responses could be made in confidence. BSP also hosted a dedicated website page where comments could also be submitted in confidence.

From BSP's knowledge of the area and initial research, it was anticipated that there might be some negativity in regard to the relocation of the post office. From discussions at the consultation event, BSP and other team members were able to discuss the concerns with local residents. Although rumours regarding the relocation of the post office circulated, BSP was able to communicate the facts to local residents and the town council, which allayed fears.

While some residents responded negatively in regard to the design of the scheme, these comments were taken on board and the design was revised before the submission of the planning application.

The planning application was submitted 28 March 2014 and was successfully granted planning permission on 4 June 2014.

Box 18.7 Case study: overcoming negative reactions

Sainsbury's Local – Whiteladies Road, Bristol

Sainsbury's took over the lease of a former Woolworths store in summer 2011. Since Woolworths closed, the building had been occupied by tenants, creating an indoor market-style environment. The site location was very suitable for a Sainsbury's Local as it is in the heart of a thriving community, with local restaurants, pubs and shops all within walking distance of nearby residential areas and student halls of residence.

The announcement, however, came shortly after the opening of a similar convenience offering by another food retailer in nearby Stokes Croft. This had caused widespread discontent among the local community, which is very active in support of independent retailers.

Given this volatile backdrop, the Sainsbury's announcement attracted significant media interest, in particular from the local BBC which used it as a platform for a piece on the proliferation of convenience stores in the Clifton area in the wake of the Stokes Croft row. In addition, a very active group of local campaigners called a public meeting to specifically discuss the possible impact on independent traders. This group chose not to accept a meeting with Sainsbury's to discuss their concerns.

Sainsbury's took a proactive approach from the outset to ensure early dialogue on the proposals with key stakeholders, to understand concerns and provide information on the store. Briefings were held with local ward councillors at which issues such as deliveries, licensing hours, disposal of commercial waste and sustainability credentials were addressed. A number of concessions were made in response to their desire for the shop to be integrated into the local retail community, for example, having an open shop front to generate a sense of being part of the community and the avoidance of 'A' boards on the narrow pavement. The opening of the store was also seen by them as an opportunity to revitalise a dormant traders' association, and it was agreed that the store manager would progress support for this once the shop was open and trading.

Letters were mailed to all residents and businesses within a quarter-mile radius of the proposed store, which resulted in a number of employment enquiries from local people.

The planning application for the store was approved and the store opened in autumn 2011. The recruitment drive attracted a significant response and many local young people and students took up employment opportunities in the store.

The purpose of a genuine, two-way public consultation in communications terms is not to unearth positive attitudes towards a proposal, but to gain maximum local involvement and hear all points of view. Negative opinion is therefore inevitable: in fact, a public consultation which reveals 100% in favour of a scheme is likely to be unconvincing, much as it may seem desirable.

Disappointing results

Consultation results may disappoint local residents for a variety of reasons, but this is of particular concern when there is a disparity between resident sentiment and consultation findings or when the consultation has not been publicised adequately or opportunities for involvement are limited. Disappointment can fuel negativity – sometimes online, sometimes in the local media – and in the case of a developer's planning application, may well coincide with the point at which the planning application is being consulted upon by the local authority or considered by the planning committee. At this stage it is generally too late for developers to adapt proposals in the light of constructive comment, and faced with possible criticism at a planning committee, the options are to withdraw and amend the application or risk it being refused.

Appropriate research and planning provides a good basis for consultation and when this is done well local residents will be involved to an appropriate level and should not have grounds to object to the form of consultation. Importantly, reference to the consultation mandate will enable the development team to negate criticisms of the process.

When effectively monitored, concerns about specific development proposals will be identified at an early stage, enabling responses to be addressed while the consultation is still live.

Managing expectations

The potential for a substantial new facility impacting on their lives and a commitment on behalf of its sponsor to consult widely can raise expectations among local residents. If not met, high expectations can lead to criticism of the process and negativity towards the proposal.

Pre-consultation can enable a developer to discuss the remit and nature of the consultation with the local authority, special interest groups and, in some cases, residents at an early stage. Where a gulf exists between expectations and reality, this should become immediately apparent and can be addressed. Often the solution need not be to offer more by way of consultation, but to consult in a way which is more suitable to the specific community.

The process of consultation should be clarified in the consultation mandate and this document made widely available to ensure that those participating understand the remit of the consultation.

A keen interest – and particularly a positive one – can be welcome news, but the consultation team should be conscious of over-promising and ultimately disappointing. Tactics should balance the need to motivate residents to secure their involvement, with tactics which will produce an appropriate level of feedback and a deliverable scheme. Sometimes the involvement of a 'middle man' as described earlier, in the form of a local authority officer (in the case of a privately led planning application), consultation manager or community arts worker, can help manage expectations.

As ever, the evaluation of the consultation will be helpful in justifying the applicant's actions: where a specific consultation framework has been put in place using pre-consultation dialogue and research, accepted by planners and run according to the consultation mandate, local authorities will understand that the consultation has met its objectives, despite any opposing voices.

Apathy and consultation fatigue

To many applicants, the prospect of an over-vociferous local audience would be very welcome: today, audiences pestered on a regular basis for customer relations feedback are more likely to ignore a consultation. Consultation fatigue is rife even among the most engaged of audiences: in researching this book, a number of planning consultants and developers interviewed admitted that they didn't always respond to consultations on development proposals in their own neighbourhood!

The issue seems to be particularly common among community and voluntary organisations which lack the resources to respond as thoroughly as they may wish. As described in Chapter 4, a range of factors can cause apathy. It is prominent in areas in which there is substantial change and therefore numerous demands on residents' time by developers. It can also result from local residents having been consulted previously but receiving little feedback or seeing no change as a result of their efforts, or where a planning application which received considerable local objection was consented at appeal. As mentioned previously, inadequate consultations on local plans can also cause local residents to resent the planning/political system.

A common problem is that those with an objection to a scheme are far more likely to take part in a consultation than are the 'silent majority'. Anger and fear are excellent motivators; a general acceptance of a project less so.

Although apathy cannot always be attributed to latent support, a case for this argument can be made in the instance of a well-publicised and -run consultation which meets objectives agreed in advance with the planning authority.

Using the strategic approach to conquer apathy

While it is very difficult to know how proposals are likely to be received, those consulting should be wary that an inadequate approach could invalidate a consultation, and pre-consultation dialogue should be used to identify the most appropriate methods of consultation.

The consultation mandate can reassure potential users of the quality of consultation, including a commitment to be open, honest and thorough and to use the results constructively.

The research stage should not end when the consultation strategy is in place, as the developer can continue to learn about the community throughout the consultation. If using online consultation, for example, it is extremely useful to use a tool such as Google Analytics to understand how people are interacting with a website: at what times of day they are most likely to comment, which pages they favour, and very importantly, which page most often leads to them leaving the website.

Presentation is very important in gaining traction for a consultation. A consultation should appear to welcome engagement and public relations has an important role to play here, yet very few consultations are designed to appeal to residents in the way that, say, advertising would.

Timing is also important. Just as the best time to put a house on the market is late spring or early autumn, this too is the best time to consult, and for the same reasons: there are no major holidays, the weather is unlikely to pose a problem and the long days allow people to venture outside in the evenings. Local authorities should be mindful of consultations being run simultaneously and possibly generating confusion. Often, local authorities have an online diary of consultations which prevents clashes from occurring.

A commitment to accessibility, discussed later in this chapter, and the possible use of a 'middle man' as mentioned earlier are also important in preventing apathy.

In some cases, such as on a residential scheme, it is simply not possible to engage with future users and can be beneficial to create a representative audience. Developing thematic or geographic panels of people or representatives can also be constructive.

It goes without saying that inspiring people with engaging and meaningful tactics is one of the best means of countering apathy. Keeping in touch following their initial involvement, thanking those who have taken part and updating them throughout the process is likely to encourage further support. In some cases, new issues can be introduced during a consultation to prevent the topic from becoming stale, although it is important not to confuse the original aims and objectives nor to introduce new ideas without giving those who have already responded an opportunity to comment.

Care should be taken to ensure that local residents' expectations are not unrealistic: it is better that many people participate to some degree than a small handful participate to a greater degree. And while high attendance figures at a planning weekend is a considerable achievement, it is also worth providing a means by which the 'time-poor' are able to take part.

Because of its accessibility, online consultation provides an opportunity to gain the voice of the silent majority or, where this is not forthcoming, gain some knowledge of their position on an issue in the absence of direct engagement: if 10 angry people attend a meeting, it is impossible to know whether the remaining local population supports or opposes the proposals, or is even aware of them. But if 10 angry people comment in an online forum and data reveals that 10,000 others visited the site, downloaded relevant documents, watched the videos but chose not to comment, then it is possible to deduce that the anger of 10 individuals is not necessarily representative of the entire community.

Finally, it is worth considering that apathy can be on the part of the resident, local authority or developer: a successful consultation requires all parties to be enthusiastically involved. The cause of consultation fatigue may well lie within the development industry, resulting from ineffective consultations which fail to deliver meaningful results. The industry as a whole has an obligation – and an opportunity – to overcome this.

Hard-to-reach groups

As earlier chapters have shown, the need to involve hard-to-reach groups is perhaps the most enduring issue in consultation. That said, the definition of those classified as hard to reach is changing.

Previously, the elderly, disabled, black and minority ethnic (BME) and women were singled out as requiring additional outreach support. Today, older age groups and women may be among those most likely to respond to a consultation, but issues still remain: whereas the 65–75 age group is very likely to contribute to a consultation, the very elderly remain unrepresented; and whereas women are now much more likely to engage in consultations independently of their husbands than they would 50 years ago, parents of young children are hard to reach due to time pressures and the practical difficulties of attending evening events. Recent entrants to the list of hard to reach are those who work, particularly commuters.

And while accessibility for commuters has been very successfully addressed through online consultation, requiring people to use IT to respond to a consultation – perhaps to be more proactive in finding the information, to be expected to do so via online networks, to comprehend information on screen and type responses – has accessibility issues in itself.

Pre-consultation research and dialogue can assist by enabling a real understanding of a local community. It can help create a picture of those groups most likely to be affected by the proposed changes and then identify those most relevant to the consultation. Local authorities consulting on strategic planning will have equalities agendas that they must comply with. Policies will recognise that different sections of the community, particularly minorities, have specific needs which should, as a democratic right, be recognised. Failure to take account of people's differences could result (particularly in the case of public bodies) in claims of indirect discrimination.

For developers, particularly those in the private sector, there are considerable benefits in reaching out to specific groups. Discussions with the local authority at pre-consultation stages should identify the ways in which the consultation may be made representative of the wider community.

It goes without saying that it requires a greater investment of time and other resources to work with those groups identified as hard to reach. Early dialogue should be used to ensure that issues can be identified prior to the consultation strategy being put in place and therefore to ensure that these groups' interests are considered throughout.

It is important never to treat hard-to-reach groups as an undifferentiated mass: each of those groups identified above (and the people who constitute them) have individual interests and needs. A consultation should have a clear understanding of those groups it needs to reach and invest time in understanding them. This might include knowing where specific groups congregate, the communication channels that they use and the issues that concern them. It is always very helpful to identify the leaders – both formal and informal – with a view to establishing initial contact through a representative. Local authorities are well placed to advise on specific groups, and in many cases can provide or make recommendations regarding translations and interpreters and advise on physical accessibility. For long-term consultations, it is often prudent to employ or train a member of staff with responsibility for specific groups.

A consultor should consider whether processes are too restrictive. For example, some consultations will only accept responses made in writing or those made at a specific event. Again at the early stages, the various ways in which responses can be elicited should be considered – always ensuring consistency with the consultation's objectives and the ability to analyse and evaluate responses and to maintain consistency throughout the process. The consultation mandate should stipulate that the principles guiding the consultation will include those of openness and accessibility, and the consultation should reflect this commitment throughout.

Box 18.8 Reaching BME groups

- Gain an understanding of the various ethnic groups within the community at an early stage
- Identify issues that concern BME communities and make the consultation relevant to them
- Explore the opportunities for training on racial awareness to help challenge stereotyping, perceptions and assumptions
- Aim to engage with BME communities by taking advantage of existing social networks, community groups, and trusted advocates and use online consultation to support, rather than replace, face-to-face interaction
- Build the leadership capacity of those who have an active interest and encourage them to use their contacts to grow the network
- Recruit BME residents onto stakeholder engagement groups
- Consider the need for translation of leaflets and the use of translators/facilitators at meetings
- Consider dietary requirements
- Avoid holding events in venues where alcohol is consumed
- Ensure that engagement does not clash with faith days
- When creating a consultation website, consider images/signs which will benefit those who do not have English as a first language, but bear in mind that not all icons and symbols translate across cultural boundaries
- Where a language other than English is widely spoken in a particular community, consider providing translations on the website – but plan how the development team will interact with those communicating in another language via online forums
- Evaluate success by asking for feedback

Box 18.9 Reaching younger people

- Avoid patronising young people
- Remember that you may have to overcome suspicion or mistrust among younger audiences. Invest time in building a relationship and demonstrate a genuine desire to hear the views of young people. Where possible, involve young people in steering the consultation, making it a consultation by young people, for young people
- Consider targeting specific subgroups based on age, ability and interest
- Investigate any existing consultative forums for young people
- Consider going to a young persons' group rather than expecting them to attend events run by the consultation team
- Work with schools and colleges, perhaps addressing an assembly, running a competition or offering a site visit
- Ensure that events are accessible via public transport and consider covering travel costs
- Embrace new technologies such as consultation websites, text voting and apps, perhaps including young people in their design
- Avoid a formal approach to events and aim to make the consultation engaging through using 'ice-breaking' games, providing refreshments and allowing ample opportunities for people to socialise
- Bring young people into the consultation by engaging them on relevant issues such as education and housing and use this as a means to involve them on more general topics
- Be aware of adults' potentially negative attitudes towards young people, which may assume lack of experience, ignorance, difficult attitudes and a lack of interest or respect
- Evaluate success by asking for feedback

Box 18.10 Case study: engaging with young people

Argent (King's Cross) Ltd and Fluid Design – King's Cross, London

Developers Argent (King's Cross) Ltd worked with Fluid Design to carry out a consultation with children and young people through their schools and youth groups.

King's Cross was known to attract young people and levels of youth crime were high in the area. Fluid approached the task as an early opportunity to tackle issues in the neighbourhood as well as to ensure that the voices of young people were taken into account.

Wide-ranging tactics were selected to address the diverse audience. These included:

- Creating a mind map
- Hot spot exercises whereby maps and cartoons were used to illustrate ideas
- Requesting that consultees complete canvas cards – A5-sized cards which included a Polaroid photograph alongside a single most important like, dislike and question. Canvas cards were displayed in prominent locations and the issues raised on the cards prompted others to complete them.
- A Youth Parliament, run by Camden Council
- Art houses, at which young people were encouraged to either complete homework or create an artwork
- An interactive matrix which allowed comments to be viewed by contributor (using a photograph) or by issue
- Video vox pops

Fluid selected attractive and interesting venues, including the St Pancras hotel and a canal boat (complete with DJ) on the Grand Union Canal as a means of encouraging young people to attend events.

The ideas generated by young people substantially impacted upon the development proposals, leading among other things to increased access east–west through the site, the retention of three gas holders on site and the creation of a fourth.

Box 18.11 Reaching older people

- Seek advice from older people on appropriate channels of communication and venues
- Carry out an accessibility audit of all consultation tactics
- Promote the consultation through libraries, post offices, churches, surgeries, hospital waiting rooms and/or bus operators to display information
- Avoid making ageist assumptions about older people and the amount of experience and expertise they may or may not have

- Be aware of any particular barriers to communication, such as language, hearing or dementia
- Ensure speakers can be heard clearly – microphones should include the hearing loop system
- Ensure that older people are represented in any stakeholder groups
- Consider that jargon and acronyms can have a disempowering impact on older people
- Ensure that print, format and content of documents is accessible
- Speak directly to the older person rather than to their carer or companion
- Evaluate success by asking for feedback

Box 18.12 Case study: involving elderly people in community development PB Planning/Barratt Homes – Pontefract

PB Planning was appointed by Barratt Homes to consult with local people in Pontefract on proposals to deliver new homes on the edge of the settlement area.

Initial research revealed that a large proportion of local residents were elderly and many would be unable to attend traditional consultation events easily. Consequently, a large proportion of the consultation was house visits: PB Planning sent letters to all residents, with the phone number and email address of the planning consultant. A series of days was allocated for the purpose and then visits booked with local residents.

The reception was very positive: local people appreciated the one-to-one contact, they were given time to understand the proposals and the impact of the proposals on their local community and were generally very supportive of the planning application.

Box 18.13 Online consultation and accessibility

- Design a well-defined, clear focus with minimal and intuitive navigation steps
- Ensure that websites are accessible with directional controllers such as D-pads, trackballs or keyboard arrows

- Allow functionality via the keyboard, rather than relying on the mouse, enabling those who use assistive technologies to access the website
- Avoid controls that change function. If these are necessary, ensure that the content descriptions are changed appropriately
- Make it easier for users to see and hear content by separating foreground from background
- Ensure that web pages appear and operate in predictable ways
- Ensure that buttons and selectable areas are of sufficient size for users to touch them easily
- Provide time for content to be read and understood
- Avoid having user interface controls that fade out or disappear after a certain amount of time
- Bear in mind that HTML is quicker, easier and more widely accessible than PDF
- Consider common forms of colour-blindness when determining colour palettes
- Ensure that text size can be increased without detriment to layout or meaning
- Ensure that the website is usable by commonly used screen readers such as JAWS, NVDA, VoiceOver for OS X, Window Eyes and Supernova and basic operating system screen magnifiers such as ZoomText and MAGic
- Ensure that the website is compatible with speech recognition software such as Dragon Naturally Speaking
- Provide content descriptions for user interface components that do not have visible text, particularly ImageButton, ImageView and CheckBox components
- Use *alt text* for important images such as diagrams and timelines, enabling those who use a screen reader to understand the images
- Where possible include standard interface controls in designs rather than custom-built controls
- Provide a text transcript of audio or visual files for people who are deaf or hard of hearing
- Evaluate success by asking for feedback

The important consideration of engaging with specific groups has been covered only very briefly here, partly because the principle of accessibility should be considered at all levels of a consultation and as such is covered elsewhere in this book. There are also many other resources online and in print which address the needs of specific groups, many of which can be found in the Further Reading section.

Box 18.14 Case study: marginalised, hard-to-reach groups and wider outreach

National Grid

National Grid is committed to ensuring that consultation processes and associated communications are made as accessible to as many parts of the community as possible. It recognises that there are individuals, groups and communities within the areas of consultation with barriers which could prevent them from fully taking part in the process.

In developing strategies to support non-statutory pre-application consultation, National Grid seeks to work with officers at relevant local authorities to identify how best to ensure consultation is inclusive and that the most appropriate methods and techniques for engaging with marginalised and hard-to-reach groups in their respective areas are used. National Grid uses this information, feedback received from the project to date and further research to create a list of the groups.

These groups often include:

1   **Geographically isolated communities:** the geography of the area that National Grid consults in is predominantly rural. Disadvantaged communities tend to lie in isolated geographical pockets. This makes it difficult for people living there to engage without access to public or private transport.
2   **Economically inactive individuals and socially deprived communities:** while individuals living in these communities may have only limited access to the internet, there is a high percentage of mobile phone ownership. Text messaging services may therefore present an alternative to email.
3   **Young people:** in addition to using social media, National Grid seeks to engage young people by making online and offline consultation more visual and interactive to reduce reliance on reading and writing.
4   **Older people:** National Grid recognises that there may be older people who have health issues or who are housebound. Those receiving domiciliary care, or in nursing homes generally, require a high level of care which makes it hard for them to attend events.
5   **Disabled people and those with learning difficulties:** this encompasses a wide range of needs and access issues, requiring a combination of consultation methods and information formats to ensure inclusion.
6   **Ethnic minorities:** ethnic groups, and their cultural and language requirements, need careful consideration.

7   **Holiday home owners, tourists and visitors:** while these are an important and distinct group, their transient nature can make them particularly hard to reach.
8   **Time-poor busy working people:** working people, particularly those who have to travel away from the area or work shifts, are a hard-to-reach group. It is important to identify those major employers whose workforce travels from across the local area and the surrounding regions to work in a central location, and where possible, seek to identify internal communication channels specific to these employers which the project team could tap into.

In engaging with these hard-to-reach/marginalised groups, National Grid understands that there are a range of organisations which act as a gateway to reaching groups with wider interests. These include those groups and organisations involved in disabled assistance, elderly care/support or promoting the interests of younger people.

Prior to launching the consultation, National Grid will contact these organisations to seek their advice on how best to reach those they represent. Based on their advice, National Grid will undertake specific activities (such as briefings or events) or make information available that best meets the needs of those they represent.

Appropriate activities include:

- Maximising the use of existing methods and networks with which people are already engaged
- Considering requests for consultation material in different languages and formats (such as large print or audio) and making them available where appropriate
- Delivering tailored presentations to representative forums and organisations to raise awareness of proposals and increase awareness and understanding of the consultation process
- Attending events that specifically target identified groups
- Using online and offline channels that specifically target the identified groups, such as community and sector-specific newsletters and websites
- Considering the best way to establish dialogue with appropriate target groups through online media
- Ensuring that a reasonable proportion of consultation exhibitions are held in venues visited by target groups
- Providing briefings and updates for relevant support agencies on the project, the consultation process and how to participate so that they are confident and able to inform and advise service users about our consultation and possible impact of the proposals

- Targeting holiday parks and second homes with tailored information to encourage seasonal visitors to register to receive project information at their home address
- Ensuring that distribution of information materials, such as project newsletters and advertisements, covers grassroots locations and community groups
- Maintaining a dialogue with organisations representing and working with the identified target groups to monitor and review the inclusivity of our consultation activity
- Siting events in locations convenient to main employment hubs to target the time-poor.

Box 18.15 Case study: maximising inclusion strategy

Atkins Global – Horizon Nuclear Plant, Anglesey

In its substantial consultation on the Horizon Nuclear Power project (Wylfa Newydd), Atkins Global put in place the Maximising Inclusion Strategy to ensure that the consultation engaged effectively with the diverse communities on Anglesey, particularly those who traditionally do not, or find it difficult to, respond to formal consultations. The planning application is an NSIP and as such the strategy sits alongside the Statement of Community Consultation (SoCC).

The key target groups for the Maximising Inclusion Strategy were identified as:

- Young people
- Older people
- Economically inactive people
- Socially deprived communities
- Disabled people and people with learning difficulties
- Minority ethnic groups
- Groups representing religions and other beliefs
- Lesbian, gay, bisexual and transgender community
- Holiday home owners.

The groups were selected in close collaboration with the Isle of Anglesey County Council. Initial meetings were held with the council's Economic Development Department and Corporate Policy Group to seek advice on the types of groups that should be targeted and the best ways to reach them. A comprehensive list of 'gatekeepers' – i.e. organisations and individuals working with and/or representing the target groups – was created.

Meetings were held with a number of gatekeeper organisations to seek advice on how Horizon might be able to utilise existing knowledge, networks and communication channels to raise awareness of its consultation within communities on Anglesey. These meetings were the first step in establishing lasting links and relationships with groups who actively support marginalised, disadvantaged or disengaged communities and individuals.

Horizon met with and sought advice from the following representative individuals/organisations:

- A project aimed at creating a culture of enterprise among young people and promoting local career and business opportunities
- An independent agency providing support and advice to voluntary and community groups
- Community Voice
- Jobcentre Plus
- Môn CF (previously Communities First)
- Social Services
- Stonewall Cymru
- The Citizens Advice Bureau
- The Council's Policy and Strategy Unit
- The Federation of Young Farmers Clubs
- The local disability forum
- The North Wales Police Diversity Officer
- The North Wales Regional Equality Network (NWREN)
- The Older People's Strategy Officer
- The Regional Community Cohesion Coordinator
- The Youth Service
- Young people's forums

The meetings held with 'gatekeeper' organisations highlighted a number of key considerations and challenges, which Horizon will take into account when planning its consultation programme:

- Getting young people interested is particularly challenging, and any efforts to consult with them must be interactive, relevant and fun, involving less reading and writing than for adults. Online activities (including social media) are increasingly becoming the most effective way to communicate and raise awareness with young people.

- Many active and mobile older people get involved with activities run under the council's Older People's Strategy and Agewell initiative. However, there are many other older people who have health problems and are housebound. Those receiving domiciliary care or who are in nursing homes generally have a high level of need and therefore it is harder for them to access consultation events.
- Around 36% of people on Anglesey are economically inactive, and there is a high level of illiteracy among adults. Furthermore, many people on Anglesey are not online. Consultation methods involving reading and writing, or requiring internet access, may therefore exclude some parts of the community. Support and guidance should be provided to assist participation.
- 2% of Isle of Anglesey's areas fall within the 10% most socially deprived in Wales, as identified by the Welsh Index of Multiple Deprivation 2011. These communities are often self-contained and disengaged.
- The term 'disabled' encompasses a vast range of needs and access issues; this group is the least homogeneous and will call for a combination of methods and information formats in order to ensure full inclusion of people with different mental and physical disabilities.
- Holiday home owners are an important group in that they spend a lot of time on Anglesey and are likely to be impacted in some way by Horizon's proposals, but they are difficult to locate and therefore particularly hard to reach.
- Whilst 98% of people on Anglesey are from a white ethnic background (almost all of whom are also identified as British), according to 2011 Census figures, the number of people from minority ethnic groups is likely to increase due to migration trends. Those from minority ethnic backgrounds are often hidden in society, and identifying cultural and language needs requires careful consideration.

Following the initial research, Horizon put in place a detailed programme of activities which specifically aimed to promote awareness of its consultation amongst potentially marginalised and excluded groups. Broadly, this involved:

- Maximising the use of existing communication methods and networks with which people are already engaged, using online and offline communication channels which specifically target the identified groups (e.g. community and sector-specific newsletters, websites).

- Ensuring that all non-technical documents are bilingual (Welsh and English) and considering requests for consultation materials in different languages and formats, and making them available where appropriate.
- Delivering tailored presentations to representative forums and organisations to raise awareness of Horizon's proposals, increase understanding of, and encourage participation in, its consultation process.
- Attending events which specifically target the identified groups.
- Considering how best to establish dialogue with appropriate target groups through online media.
- Ensuring that a reasonable proportion of consultation exhibitions and drop-in surgeries are held in venues visited by target groups.
- Where appropriate, minimising the use of technical language and jargon in written consultation materials to ensure that basic messages reach all audiences regardless of age and literacy levels.
- Providing briefings and updates for relevant support agencies on the proposals, the consultation process and how to participate so that they are confident and able to inform and advise service users about the Horizon consultation and possible impacts of the development.
- Targeting holiday parks/second homes with tailored information to encourage seasonal visitors to sign up for future information about the consultation at their home address.
- Providing a staffed, dual language (English and Welsh) telephone helpline to deal with any queries about the project and the consultation process and give guidance on submitting feedback.
- Ensuring distribution of information materials (e.g. Horizon newsletter) and consultation advertisements covers grassroots locations and community groups.
- Ensuring that consultation venues are fully accessible, and that events are held in locations which are regularly attended by marginalised groups – including people in rural communities.
- Maintaining an ongoing dialogue with organisations representing and working with the identified target groups to monitor and review the inclusivity of engagement and consultation efforts.

Online consultation and perceived risk

This chapter has so far identified external issues which may affect the smooth running of a consultation. Concerns about online consultation which then threaten the effective use of this powerful tactic, however, is one which may exist in the consultation team itself.

Online consultation is relatively new, and a fear of the unknown persists despite many success stories. As Chapter 16 identified, the consultation scene was altered radically with the introduction of online consultation impacting on the share of power, the geographic spread and a new set of consultation tactics. Online consultation has brought about numerous benefits – engaging tactics, increased accessibility and thorough reporting to name but a few – but those running consultations frequently worry that the consultation will be hijacked by trolls, that the website will be open to corruption and that registration will be off-putting.

Online, a 'troll' is an entity which takes part in discussions purely to disturb other users. Potentially, a troll can anger people, disrupt the flow of debate/discussion and use abusive language. Anyone who has run a consultation will know that this behaviour operates both on- and offline. Online, there can be effective means of dealing with trolls. However, it is extremely important to identify this as either anti-social behaviour or merely an impassioned and negative response to the consultation: unpalatable though it may be, the latter should not be dismissed, as everyone is entitled to put forward their views on the subject being discussed. However, activities which are clearly anti-social and unconstructive can be stopped if the consultation mandate set out the basis upon which people are invited to respond and specific rules and regulations in relation to harassment, bullying and bad language are put in place (perhaps contained within a user guide[4]). Software can be used to identify bad language and 'spam' and is advised, in conjunction with monitoring. Where necessary, posts can be removed with immediate effect, IP addresses banned and usernames invalidated. Where a local issue has potential to escalate into a national issue and draw response from across the globe, mechanisms can be implemented to allow only those within a specific postcode area to register to take part in online forums, and where necessary the electoral register can be used to check the veracity of identities given. Where the consulting body has control of the website on which such activity occurs, preventative action can be almost too easy: it should only be used when absolutely necessary.

With hacking, phishing and spam affecting our daily lives, it is unsurprising that issues of cyber security concern those running consultations. Certainly, an unprotected website can leave itself open to abuse, and where user details are being collated via an online database, the legal and reputational impact can be considerable. However, all websites can benefit from EV (Extended Validation) SSL (Secure Sockets Layer) certificates. Websites with this functionality display a padlock icon and the https (Hyper Text Transfer Protocol Secure), rather than simply http (Hyper Text Transfer Protocol) in the URL. This means that all communications – including user names and passwords – between the browser and the website are encrypted and only accessible by the website owner.

As mentioned previously, registration can be extremely beneficial in restricting consultation responses to a specific locality and understanding more about those taking part. However, those running the consultation should also consider the downsides of registration: potential users may be reluctant to pass on email addresses, passwords and other personal data, and may be put off by the amount of time (perceived or otherwise) that registration demands. Consultation websites should seek to make the process simple and reassuring, explaining the need for registration, referring as appropriate to the security measures in place and making the process as smooth and simple as possible. Typically, a consultation website will require a name, postcode, username and password. Any other information, such as a full postal address or demographic data (age, employment or marital status) should be given voluntarily, and it is advisable to request this data at a later stage, allowing the user to have built up trust and respect for the consultation and appreciate the benefit to the consultation in supplying such data.

Understandably, tactics as new and as powerful as online consultation can raise concerns. But by far the greatest risk in online consultation is not connected to the consultation website itself but the absence of it: failure to provide a platform by which local residents can discuss a proposed development online can result in the developer being unaware of other online discussions, which can then gather momentum and perhaps only come to light when it is too late to address concerns or misapprehensions.

Negative media involvement

There is an assumption that the media, specifically local newspapers, are naturally anti-development, that a newspaper will always champion the voice of the local resident over that of a corporate entity, and that bad news is more likely to make the headlines than good news. There is some truth in this, but this does not justify developers failing to engage with journalists.

As with local residents, positive relationships with the media are based on provision of information and a positive, open and transparent approach.

A shocking proportion of developers opt not to communicate with the local media in the early stages of consultation, entering into dialogue only (and often reluctantly) when a negative issue has been brought to the attention of the media. Frequently a negative, unbalanced and perhaps inaccurate story will have been published by this stage, causing substantial damage both to the consultation and the reputation of its partners more generally.

The recommended approach is to contact the local newspaper at the early stages of the consultation: use the consultation mandate to explain the process and remit of the consultation, ensure that the local media is fully furnished with the facts and the positive messages and has contact details for an appropriate individual in the case of future questions. The result of this approach is typically a positive story in the first instance, and a more balanced story should local residents approach the newspaper with concerns about the consultation or development proposals. The local newspaper can also be used to publicise consultation events both in print and online.

Box 18.16 Case study: positive media relations

Bayfordbury Estates – Brookfield Riverside, Hertfordshire

Brookfield Riverside was one of the first schemes to use social media in consultation, using TTA Group (part of Chime plc) and later PNPR.

In 2008, Bayfordbury Estates consulted local residents on the outline proposals for Brookfield Riverside, a mixed-use, predominately retail scheme. It did so through a variety of methods and chose to use Facebook as a means to reach young people and commuters.

The first person to engage on what was possibly the first Facebook page for a planning application was a journalist from the local newspaper, Gemma Gardner. Gemma was invited to Bayfordbury Estates' offices and shown the Facebook page at the point when it went live. She then became the first person to post on the Facebook page, marking what continued to be a very constructive relationship between the developer and the local media.

Commenting on the initiative, Gemma said,

> I was pleased to be shown the Facebook page and to use it at an early stage in the consultation. At that time using social media to engage with the public was rare locally. I was impressed by the lengths that the developer went to in reaching out to local people at the start of their bid and I was in touch with the developer throughout the planning process, which I believe benefited the local residents and the planning application.

Albeit a one-way tactic, local media relations represents an excellent opportunity to communicate with a wide audience. And thanks to the proliferation of local newspaper websites (now more numerous than those newspapers producing a print version) this is changing: opportunities exist to drive readers to the consultation website, or to encourage discussion via a local newspaper blog or social media page and in doing so a once static, asymmetrical means of communication becomes an interactive tool.

Conclusion

This chapter has addressed some of the most significant issues affecting today's consultations. The enduring answer to these and other issues is the strategic approach to a consultation: with an appropriate level of pre-consultation dialogue, stakeholder research, situational analysis, a clear strategy communicated via a consultation mandate and an appropriate selection of tactics, issues can be mitigated.

There is no means of prescribing a risk-free consultation, because no such thing exists. A consultation which is open, transparent, accessible and two-way will encourage both positive and negative comment, as is expected. A good consultation is not one which is free of negative comment, but produces informed responses which are constructive in shaping future plans and are used accordingly.

Notes

1 An abbreviation for Not In My Back Yard, NIMBY is the pejorative term used to describe those who oppose a development because of its proximity and impact on their own home, even where they would recognise its value elsewhere, on grounds that are selfish and subjective. The authors of this book do not support the use of the term, believing that individuals have a right to express concern about a potential development and that the planning system is well placed to determine whether that view is a relevant planning consideration.
2 Rydin, Y. (2011) *The Purpose of Planning*. Bristol: Policy Press.
3 CBI (2014) Building Trust: Making the Public Case for Infrastructure. London: CBI.
4 A template User Guide is shown in Appendix 4.

Part IV
Post planning

19 Community relations during construction

Introduction

With the pre-planning consultation complete and planning consent won, a significant amount of community engagement has been accomplished. But as the project moves into the construction phase, community engagement too enters a new phase.

A successful pre-planning consultation will have created a good foundation for the next stage of engagement. But rather than a continuation of the work to date, community engagement post-planning has new aims and objectives, new stakeholders and new challenges. The development team will change with the addition of contractors and perhaps the omission of the planning consultant; new teams within the local authority will have an interest and the local dynamic will metamorphose as residents/users of the new development emerge. A proactive approach towards community relations is required to ensure a constructive relationship between all parties.

Why invest in community relations?

A construction management plan (CMP) is a set of conditions put in place by a local authority to ensure that developers minimise the negative impact of construction on the surrounding community. A CMP typically includes restrictions on working hours, vehicle movement and access, parking and loading arrangements, parking bay suspension, temporary traffic management orders, the impact of scaffolding on public highways, the height and aesthetics of hoardings, how pedestrian and cyclist safety will be maintained, control of dirt and dust, waste management and communication with local business and residents. Planning consent rests on the CMP, and if the developer fails to comply with it, development can be halted. Therefore it is vitally important that the developer upholds the CMP. For larger projects, an individual is often put in place to oversee liaison between the developer and the community to ensure smooth communication where possible and to gain an understanding of any problems that might arise.

Good community relations can also protect and enhance the reputation of the developer and other members of the project team. Without fear of generalising, it can be said that the process of construction is rarely popular. The imposition of a highway across previously unspoilt countryside, the doubling of a small town with a new housing estate, or the construction of a power facility – few such changes to the physical landscape are welcomed. Where negative sentiment already exists within a local community, a developer has an uphill struggle to deliver a project while maintaining a good reputation. And once work begins, that reputation can be further tarnished due to frequent movements by construction vehicles, the noise of pile driving, road closures, parking cessations, occasional cuts to power supplies and numerous other, often unpredictable, consequences of construction. Managed badly, this can lead to protests, negative media coverage and a torrent of criticism online. So the challenge to protect the development team's reputation is a complex one and has implications both at a local and at a corporate level.

If managed well, good community relations is a substantial investment in an end product. The first residents of a new housing development, if appreciative of the development team, can become ambassadors for the scheme and thus benefit sales. A business district which maintains a positive relationship with tenants despite construction work is likely to retain those tenants. And a clean and accessible shopping centre is a popular one.

For larger projects, planning permission is achieved in stages and so remains ongoing during the first phases of construction. Thus, community relations strategy will run in parallel with planning consultations. Developers of housing or mixed-use schemes frequently find that the first occupiers can be fiercely resistant to further development, often because they resent the likely impact of further development (such as development on green fields or additional parking pressures), and this is particularly likely if combined with the disruption that construction invariably brings.

The Public Services (Social Value) Act introduced in 2013 requires those who commission public services to consider the social, economic and environmental benefits that can be achieved during the construction process. It is designed to generate social value and encourage discussion between developers and local residents, with the aim of finding new and innovative solutions to difficult problems. The Act put in place the Social Value Awards (which recognise and celebrate good practice), advice on measurement and evaluation, and tools and resources to help organisations measure their impact. Although not currently relevant to private sector developers building privately funded schemes, the principles of the Social Value Act are beginning to filter through and may become more universally applicable in future.

The moral obligation to 'do the right thing' also motivates good community relations. And where a large-scale rail or highways project is being driven through communities at a significant cost to the taxpayer, local residents are also an indirect client, giving the developer a dual obligation to act responsibly and reduce the negative impact of construction.

A good relationship with the local community will not only enable the development team to minimise disruption where possible, but also to pre-empt any future problems: regular dialogue with residents can identify problems before they occur and prevent, rather than mitigate, their effects.

A strategic approach to community relations

Like the pre-planning community engagement which preceded it, good community relations is symmetrical, responsive, engaging, genuine, timely, informative and socially aware. Community relations strategy will also be guided by the corporate objectives of an organisation.

As with consultation, a community relations strategy should begin with local dialogue. Meetings with those local residents most affected and stakeholder groups representing the wider area will enable the developer to understand local residents' fears and expectations and put in place channels of communication for the future. While previous research is a useful starting point, the developer must be cognisant that interested parties may change at this stage, especially with the addition of new users and occupiers.

Box 19.1 Case study: stakeholder research

Crossrail

Crossrail is a 118-kilometre (73-mile) railway line which will provide a high-frequency passenger service linking Heathrow and parts of Berkshire and Buckinghamshire, via central London, to Essex and South East London. It gained parliamentary approval in July 2008 with the passing of the Crossrail Act.

Crossrail's community relations is guided by thorough opinion research with its various stakeholder groups. This research informs communications, ensuring the most appropriate information in the most appropriate format is provided to the right people.

Crossrail aims to communicate with all residents and businesses within a 200-metre radius of each of its potentially disruptive construction activities, often occurring at Crossrail's 40 stations but also including other worksites. The project has reached tens of thousands of neighbouring stakeholders in this way.

The starting point is demographic research, which can help paint a profile of an area, but due to its very general nature must be complimented with substantial face-to-face contact. The Crossrail team values direct contact highly, using events and house-to-house visits to build up a full image of the local residents, their concerns and their expectations.

Details of those who have expressed an interest in keeping in touch with Crossrail, including their specific interest (whether in relation to planning, construction or as future users) are logged on a huge database which enables individuals to be targeted more directly. The database is updated following every contact made and with information about the potential impact of construction on specific areas.

The 22 local authorities on the Crossrail route are also an important audience, holding responsibility for road closures, planning consent and noise levels. Crossrail seeks to provide local authority officers and members with ample information and confidence to enable them to facilitate the development.

Following stakeholder research and situational analysis,[1] a strategy for community relations during construction is created. Just as with the consultation strategy, the community relations strategy should be informed by aims and objectives for communication with local residents, clearly defined messages, legal requirements, resourcing, subjects for discussion and opportunities to involve local people. Procedures should be put in place for monitoring feedback, updating the stakeholder database and evaluating the success of the strategy on a continual basis.

Box 19.2 Case study: evaluation criteria

Crossrail

Crossrail has put in place specific performance assurance criteria, both qualitative and quantitative, upon which its contractors' community relations activity is assessed:

- Strategic planning
- Timely provision of information
- Quality provision of information
- Timely response to complaints
- Quality inputs into forums
- Implementation of community investment
- Legacy from community investment and Young Crossrail
- Community relations resourcing
- Quality of response to complaints
- Community relations integrated
- Active stakeholder engagement
- Considerate Constructors Scheme

- Submission of reports
- Planning
  - Integration of community relations function within project
  - Liaison plan
  - Rolling look ahead
- Communications
  - Timely provision of information
  - Quality provision of information
- Complaints handling
  - Timely response to complaints
  - Quality response to complaints
- Community investment

Each of the criteria is evaluated every 8–10 months to inform Crossrail of its contractors' ability to communicate effectively and to identify areas where support is required. Best practice is also shared to ensure things that work are adopted by as many contractors as possible.

Crossrail's contractors have embraced the process and Crossrail is confident that it has led to a genuine improvement in performance. The results across all areas assessed improved significantly over the first three rounds, with a plateauing and slight improvement in the fourth and fifth rounds. These results were expected, demonstrating that contracts had reached a level of maturity.

Good community relations is both proactive and reactive and is not limited only to mitigating the impact of construction. The community relations strategy for a medium or large scheme might also include outreach activities, perhaps involving education, the environment, art and employment initiatives. This proactive approach is a positive means of reaching a local audience, engaging them in the project through relevant and appealing tactics, and promoting the wider values of the organisations involved.

Opportunities to engage with local residents

Education and employment

Development teams frequently use education as a means of reaching stakeholders both during the planning stages and beyond. Working with schools reaches not just children, but their families too, often within a very specific geographic area. Similarly, developers and construction companies have skills which can be of use to the wider community and this too can help develop positive relationships with the site's neighbours.

Through these initiatives, the development team is able to better understand its local community and in doing so, address local issues, develop community cohesion, gain feedback on details within the scheme and create interest in the development – attracting workers, shoppers and residents. By investing in education and employment initiatives, the developer is also able to assist with broader economic, environmental and social regeneration.

Box 19.3 Case study: a contractor's role in creating employment and training opportunities

British Land – Clarges, Mayfair

Clarges Mayfair is a landmark residential development in the heart of London, overlooking Green Park and Buckingham Palace. British Land and main contractor Laing O'Rourke were determined to make sure that the project's economic and social benefits reached those in less affluent local areas. Jointly, the two companies developed a strategy to create employment, education and training opportunities for local residents and to support local communities in Westminster.

In the first 18 months of construction, more than 40 long-term unemployed local residents, young jobseekers and homeless women benefited from pre-employment training initiatives at Clarges, working with partners including The Prince's Trust, City of Westminster College and the Marylebone Project.

The Clarges team also hosted work placements for almost 60 young people from Westminster and neighbouring boroughs, through the Construction Youth Trust's Budding Brunels programme, which British Land supports. Placements include tours of the site, followed by challenges for the students, who are also given career advice by professionals. The young people welcomed the chance to interact with volunteers from Laing O'Rourke, British Land and other partners.

Many of the Clarges team donated time, skills and expertise to assist local charities and social enterprises, for instance providing mentoring support to Bounce Back, a charity that trains and employs ex-offenders from Brixton prison and working with local community centre, The Abbey Centre in Westminster, to create a small garden for older people to grow their own food.

Together with Westminster charity Pursuing Independent Paths, another British Land community partner, the Clarges team is also creating long-term placements for adults with learning difficulties.

Box 19.4 Case study: construction companies and education

Kier Construction

Kier Construction in Southampton secured £5,000 of funding from The Kier Foundation to help a local preschool which desperately needed its outdoor space renovated. The preschool is in one of Southampton's most deprived areas, where between 24% and 37% suffer from high levels of child poverty. The team supplied and fitted artificial turf, climbing equipment, a Wendy house and a variety of other outdoor activity equipment.

The team held fundraising events to raise further funds to support the project and has engaged its supply chain to be on hand to help as well.

Environmental

In recognition that a new development may have, albeit only in the eyes of a few, a negative impact on the fabric of a neighbourhood, developers frequently make environmental improvements to a neighbourhood. Typically this involves developing a nature reserve, creating the means by which endangered species can be protected (bat boxes are common) or making a contribution to a local park or woodland.

Box 19.5 Fostering good community relations through environmental initiatives

David Wilson Homes – Pickering, Yorkshire

David Wilson Homes (a division of Barratt plc) sought to deliver 96 homes and a new community park in Pickering, North Yorkshire.

A park was a main feature of the proposals but rather than the park being designed in detail, its features and layout were subject to local engagement.

Meetings were held with the parish council and local ward members and additionally Paul Butler (now of PB Planning and previously the planning manager at David Wilson Homes) and Peter Morris (development director) visited local schools in the area, delivered assemblies on the subject of planning a park and then held workshops with the children.

The workshops produced a wealth of unique ideas which were then drawn up by the development team and presented to the local residents. Furthermore, the parish council was actively encouraged to become the owners of the park in the future and take direct involvement in planting and other maintenance activities on an ongoing basis. Importantly, the aim was to create a framework for development of the park. This meant that David Wilson would deliver the initial infrastructure and the parish council could then decide in the future the types of uses and facilities to be included, with reference of course to the number of exciting ideas provided by local school children. This gave the local community a sense of ownership in what was a significant change for the town. The park is due to become an extremely popular feature and will benefit community relations post-construction of the homes. Early engagement was key to its success.

The arts

Using the arts as a form of community engagement enables the community to work collaboratively on a process which is creative, fun and can be directly relevant to the development itself. It can provide a positive experience in the process and result in a product which endures and provides a long-term reminder of the collaboration. The involvement of a community arts worker or professional artist can provide a helpful bridge between the developer and the community, and the process can create a sense of ownership in the new development.

Arts work can take various forms: encompassing visual and performance arts, permanent or temporary, and a product by or for the community. The resulting piece is often inspired by the architecture of the new development, or may link to the site's previous use.

**Box 19.6 Case study: use of the arts
in community relations**

**Linden Homes, East London Community Land Trust and
Alison Turnbull Associates – St. Clements, London E3**

Alison Turnbull Associates worked with Applecart Live and zURBS on 'Transitions and Connections', a public art project for a new development located on a historic site. Rather than installing a piece of professionally designed and produced artwork, community engagement was considered important to give the final artwork credibility and relevance.

'Transitions' refers to the changing uses of the site, originally an asylum and mental health hospital. 'Connections' refers to the new physical connectivity and integration of new neighbours into the community.

The engagement activities sought to collate the histories and memories of the local residents. They began with 'Stories from Home': storytelling activities with St Paul's Way Trust School, Queen Victoria Seamans Rest, Bangladeshi Women's Community Group and John Tucker House (sheltered housing).

The second stage was a series of neighbourhood walks organised by the workshop leaders. While walking, the group found objects and took photographs which were used as the starting point for model making. The images and models were then used in workshops to facilitate conversations about the area, which enabled both the development team and local residents to discover more about the people and the place.

A local artist blacksmith, Agnes Jones, joined the walking groups and designed a gate and other permanent pieces inspired by architectural elements found on retained listed buildings (ornate key stone and the leaf motifs) and icons relating to local industries (canal boats, bicycles, hammer and tongs, computer circuits and a needle and thread).

An exhibition of the designs, along with the models and film, enabled local residents to vote on those images they wished to see included in the final artworks.

Following completion of the project, the outcomes of the engagement and an interview with Agnes were filmed and made available online.

An arts strategy usually arises out of consultation with the planning department and consideration of Section 106 payments being made elsewhere.

Box 19.7 Case study: bridging the gap between show home and public art

Countryside Properties – Great Kneighton, Trumpington, Cambridge

When Countryside Properties completed Abode, the second phase of new homes at Great Kneighton, in January 2015, they were keen not only to attract potential purchasers but to invest in community relations.

This led to the unique idea of 'A Showground for Real Living': a public art project that offered residents the opportunity to reflect, imagine and build the future of their community.

Two artists spent a year living as 'residents in residence' in the three-bedroom show home. Together with other residents, they extended the show home's domestic space into a public one. The ground floor was transformed into a Showground Library and Real Living Café in order to host formal and informal discussions about the new community and health centre. Many neighbours wondered how the new facilities would take shape – physically and organisationally – so the residents set about modelling an imagined community centre in the show home.

The show home started to slowly transform into a 'showground of real living': a show-and-tell home of public life and a test site for community encounters. One of the questions that emerged was 'which voices are not often publicly heard around here?' This led the artists to invite a group of young people from different parts of the area to spend time together sharing ideas about 'real living'. They developed ideas for Trumpington Show Reals, a series of short films in response to their experience of living in Trumpington, which they then wrote, directed, shot and acted in.

The idea of a show homes pub was initiated, providing a very popular attraction on a monthly basis. Chickens joined the artists, who shared their skills in caring for them along with their gardening expertise. The chickens drew yet more people to the house, those who came to feed them or just stopped to look at them.

The show home constantly evolved into a community residence. Hosting duties moved between local residents and many of those living on the development initiated ideas. These include a neighbourly exchange of skills and services, a meeting space and a café. Importantly, the space remained fluid, enabling people to try out things.

The artists, the developers and the community reached the conclusion that community ownership is about more than giving people a voice; it is about giving people a role – in fact, it is most successful when people take on a role themselves.

When the artists departed, Countryside Properties kept the space open for public use until the scheduled opening of the new community centre, with neighbours managing it on a day-to-day basis.

Box 19.8 Case study: public arts strategy

North West Cambridge

The North West Cambridge Development, a 150-hectare mixed-use site owned by the University of Cambridge, has successfully implemented a strategic approach to public art.

A strategic approach

A number of principles inform the strategy:

- The provision of sustainable and diverse opportunities for existing and future communities to engage and benefit from the public art programme, contributing to social cohesion and community building.
- The establishment of public art commissioning as an integrated part of the development. Artists' work will be underpinned by a thorough understanding and commitment to the development context – across physical, social and historic agendas.
- The strategy approach recognises the shifting nature of the public realm and urban fabric of the proposed development over the 15–20 year development programme, offering a flexible methodology that allows for adaptation over time.
- The strategy is grounded in a practical and deliverable methodology that includes clear guidelines for delivery, best practice models for commissioning, making capital budgets work effectively and productive partnership working.

A number of strategic themes drawn from the context and character of the development inform the public art strategy:

- Twenty-first-century communities
- History and archaeology
- Ecology
- Sustainability
- Community engagement

The commissioning methodology intends to deliver a model of best practice, responding to contemporary directions in public art that emphasise research and exploration of place to produce context-specific approaches and generate genuine dialogue between artists and communities. A consideration of public art at the early stages of the development process meant that artists' contributions can be meaningfully and sustainably integrated within the physical infrastructure and as importantly across the communities that will be a part of the development's future.

The public art strategy is comprised of a series of programmatic strands that will be delivered throughout the life of the development.

1   Habitation – artist research and residency programme
2   A distributed collection – public art commissions
3   Activation – temporary and event programme
4   Making place – naming commission
5   Education programme
6   Legacy

Delivery

To deliver a long-term strategy successfully it is essential to establish a delivery structure that maintains momentum over the full duration of the construction stages and that has the flexibility to respond to the changing nature of the site. The strategy will be delivered through a three-tiered structure – the university as project owner, an internal public art advisory panel to provide high-level guidance and reporting, and an art advisor appointed to deliver all elements of the strategy.

Conclusion

The first three years of the public art strategy have seen eight residency artists working with departments and communities. Seminars, screenings, events, lectures and workshops have supplemented the artworks, providing opportunities for local communities to get involved, in addition to an archive of art material on a dedicated website showcasing the works.

By undertaking research-based and community-enhanced approaches to art, the character of the site is being created organically and helping to enrich the development process to inform the new place.

Community buildings

Community buildings are an important feature of a successful community and therefore a significant benefit in community relations. Retail and leisure schemes often provide community spaces, such as amphitheatres or space for pop-up exhibitions and markets. New supermarkets increasingly include a community room, and mixed-use developments are likely to include a community facility, be it a specific community centre, a school, sports club or church. Economic viability often dictates that a community facility is not among the first buildings to be constructed, and therefore a temporary community facility is often provided. This can provide a base for the community liaison offer and a venue for community relations activities.

Box 19.9 Case study: community building

David Lock Associates/Bee Bee Developments – Priors Hall, Corby

Priors Hall is a new community of 5,100 dwellings and associated mixed-use development on the eastern edge of Corby. The Priors Hall outline planning application, prepared by David Lock Associates (DLA), was approved by Corby Borough Council in 2007. As a catalyst for Corby's regeneration strategy, the development is a new sustainable community within an extensive parkland setting. DLA, in the capacity of 'town architect', was committed to helping deliver an exceptional scheme.

The development team recognised at an early stage that a community worker would bring about substantial benefits in integrating the new community with the town and provide a positive form of engagement with residents of the scheme.

A community worker was appointed at an early stage and allocated a new house on the development, from which she was able to coordinate a wide variety of events, such as egg hunts, barbeques and fireworks. Doing so significantly benefited community relations during construction. The physical location also vastly benefited consultation with new local residents on emerging proposals for the scheme.

Local supply chain

As part of a commitment to the area in which they are working, many developers choose to work with local contractors where possible, thus benefiting the local economy and forging links with the local business community.

Typically,[2] a local supply chain might expect suppliers to comply on a range of issues, such as:

- Health and well-being – in relation to health and safety standards
- Labour practices – respecting those they employ and offering a safe workplace, free from discrimination, harm, intimidation, harassment or fear

- Biodiversity – aiming to enhance net biodiversity and to understand biodiversity risks
- Community engagement – engaging with the local community, minimising disruption and actively supporting local projects
- Environmental management – putting in place effective environmental management systems appropriate for the nature and scale of their business and services
- Energy and carbon – where appropriate, improving energy efficiency and optimising low-carbon energy supply, taking full advantage of government initiatives, to reduce carbon footprint where possible
- Materials and waste – aiming to send zero waste to landfill, measuring and reducing waste at source, reusing and recycling; using secondary (reused or recycled) materials wherever possible
- Local employment and procurement – offering local businesses and local people the opportunity to tender for work
- Skills and capacity building – supporting apprenticeships, training and education
- Prompt payment – payment within agreed terms
- Fair practices – avoidance of corruption, bribery or unfair, anticompetitive actions.

Box 19.10 Case study: a local charter

British Land – St Stephen's Shopping Centre, Hull

British Land has increased its focus on local procurement and is encouraging its property teams to 'shop local'.

This is based on the belief that local suppliers often offer better service levels and value for occupiers and supports British Land's focus on benefiting local economies and communities.

At St Stephen's Shopping Centre, 51 contracts worth £123,000 have been won by local businesses. Hull-based firms include alarm services and sprinkler maintenance, car park tickets, cleaning products suppliers and a painting contractor. A local firm was also awarded the generator maintenance contract. In each case, the local provider offered excellent value, delivering cost savings and faster response times.

St Stephen's has achieved success in local procurement by working with facilities management partner Incentive FM to actively encourage local firms to tender and put local procurement on the agenda at quarterly meetings.

To further build on these successes, the centre manager is working with the local business improvement district (BID), Hull and Humber Chamber of Commerce and the local enterprise partnership (LEP) to promote a 'shop local' approach to business members.

Sponsorship, support and sponsorship-in-kind

The variety of skills that make up the construction and development team have a great deal to offer the local community. A seemingly simple activity, such as providing the use of the landscaping team to overhaul a pocket park or members of the construction team to rebuild a brick wall at a local school, will present an excellent opportunity to forge links with the local community. Sometimes this might involve skills sharing – teaching jobless young people the skills of gardening or brick-laying or offering talks about careers in construction at a further education college.

Sponsorship too is popular. A developer or construction company will often provide a new kit for a local sports team, which provides an opportunity both to meet local residents in a non-adversarial context and to gain brand recognition at football matches and in the local media.

Box 19.11 Case study: charitable donations

Kier Living

Kier Living's affordable housing business donated £1,000 to a local community centre, Hollymoor in Birmingham, which funded the kitchen equipment needed to open a new café.

The community centre is run by Longbridge Childcare Strategy Group, a charity which manages six childcare venues within the Birmingham area, as well as a preschool nursery at the site. The group is encouraging the local community to utilise the facilities they are providing, including the newly installed café.

The café, along with room hire and meeting space, is funding the community-based not-for-profit organisation and aims to enrich the lives of children by creating quality affordable childcare.

Site management

It goes without saying that local residents will be more positively engaged with a development team and less critical of it if their locality is kept clean and safe, and if they are provided with timely and adequate information should their daily lives be disrupted.

The following tactics are all regarded as good practice in overseeing community relations:

- The appointment of a community liaison officer is an excellent starting point, as this ensures a single point of contact for local residents, plus a coordinated and consistent approach.

Box 19.12 Case study: use of community liaison officers

Crossrail

Crossrail employs approximately 20 community relations officers directly and also works with as many community relations representatives, who are employed directly by its contractors. Both roles provide a vital link between the local communities and the construction teams.

Each community relations representative is appointed as soon as contractors begin on site and remains in place until the contractor leaves.

The contractors' community relations representatives (CRR) work on the front line, responding to enquiries and working with contractors to minimise the impact of the work on local residents. Members of the Crossrail team are in place to provide coordination between contractors and respond to complaints and enquiries which may span more than one construction site.

Stakeholder research enables the Crossrail team to identify those affected – usually within a 100-metre radius of any development site. Communication in the early stages is by letter and email. Once work begins on site, most communication is face to face, comprising formal and informal individual and group meetings. Community liaison panels, chaired by a local authority officer or member, take place in relation to each construction site. Community liaison panels follow a formal structure and are minuted.

Ad hoc group drop-in sessions are also held as required. Exhibition boards provide information, and local residents are sometimes invited to attend site visits.

Crossrail produces information sheets in relation to each station. These provide specific information about the programme of work and any activity likely to impact on local residents. Station-based newsletters are also produced to give regular updates, and Crossrail produces a corporate newsletter entitled *Moving Ahead* every three months.

The wider Crossrail community relations team meets twice yearly. In addition to the Crossrail employees and contractor representatives, Network Rail and London Underground's community relations personnel attend the meetings, at which project updates and best practice are shared through presentations and workshops.

In some cases, this role may be taken on by a construction impacts group or development forum.

Box 19.13 Case study: a construction impacts group

Argent LLP – King's Cross, London

Chaired independently, the King's Cross Construction Impacts Group comprises local residents, participants from each of the major developments in the King's Cross area and representatives from the local authorities and other major agencies.

The group meets quarterly as a minimum, or as required, according to events or issues of particular concern. The purpose of the group is to:

- Ensure that the problems and opportunities arising from the prolonged period of major construction in the King's Cross area are anticipated and considered at senior levels across the range of agencies involved
- Ensure that the responses to problems and issues are delivered effectively by the agencies and organisations concerned
- Ensure that the various opportunities arising from construction activity are fully taken up, such as employment and training, enhanced community safety and cultural activities
- Ensure that the necessary coordination between stakeholders is maintained, at the appropriate levels
- Receive briefings and updates from across the range of activities and remits and to make key, strategic decisions to act on and resolve issues as and where necessary, before they arise
- Consider matters including pollution and environmental impacts, wider health issues, traffic management, community safety and policing arrangements, effective communication, and local employment and training, in addition to other relevant activities in the area arising from regeneration programmes, private development etc.
- Look ahead at significant construction events and plan for them accordingly.

- Newsletters, emails, a community relations website and social media, telephone helplines and exhibitions in local community centres have been found to be useful in imparting information.
- Face-to-face and small community group meetings enable the development/construction team to speak directly with those individuals affected and respond to their concerns.
- Community liaison panels are a more formal means by which the development team can understand residents' concerns, but are smaller and more manageable than public meetings.

Box 19.14 Case study: community relations website

Essential Living – Berkshire House, Maidenhead

Essential Living carried out a comprehensive planning consultation for the regeneration of Berkshire House in Maidenhead, using ConsultOnline to run an online consultation alongside an offline consultation run by Forty Shillings.

When planning consent was granted and construction commenced on site, a community relations website was put in place which provided substantial information about the scheme, its design, its sustainability features, the site's history and the team. Frequently Asked Questions were posted on the website and local residents encouraged to post their own questions. Construction updates, with images of the development in progress, were posted monthly and as required, and users were invited to register both for construction updates via email and for future lettings availability. Quick links on each page provided help with navigation, while links to Facebook, Twitter and Google+ encouraged users to share information.

The website (Images 19.1–19.4) was promoted using the social media accounts set up for the planning consultation and the website monitored using Google Analytics.

Image 19.1 Berkshire House community relations website: home page

FAQS

General

Q: How have local residents responded to your proposals?

A: n Spring 2013 we carried out a comprehensive consultation which included meetings with local groups, a leaflet delivered to 10,000 homes, a two day public exhibition, a website and social media. Of the 300 that responded, 92% supported the regeneration of Berkshire House; 87% supported the conversion of the building from office into homes and 70% supported the initial design concepts that had been worked up for the building.

Q: What is the local council's view on the regeneration of Berkshire House?

A: In December 2013, the Planning Committee of the Royal Borough of Windsor and Maidenhead voted in favour of the scheme with a 10 to one majority. The scheme received detailed planning consent with conditions around detail design and materials.

Q: What kind of homes are being provided?

A: The re-design of Berkshire House will provide 68 high quality studio, one and two bedroom apartments for the rental market.

Q: How much commercial space is being provided?

A: The 5,702sqft of commercial space at ground floor level will be retained.

Q: How much parking is planned?

A: 30 car parking spaces are being provided.

Q: I am interested in renting a property at Berkshire House – who should I contact?

Essential Living

Image 19.2 Berkshire House: frequently asked questions

CONSTRUCTION UPDATES

Construction update 29 March 2016

Since our last construction update, the steelwork has been completed on floors 11, 12 and 13. Floor 14 is due for completion very soon, along...

Read full article

Construction update 26 February 2016

Construction work at Berkshire House is progressing very well. Since our last update two weeks ago, we've carried out the following work: Steelwork for levels...

Read full article

Construction update 15 February 2016

The strengthening work to the columns on the 10th floor and the demolition of the 1st floor Queen Street, High Street and Park Street external...

Read full article

Essential Living

Image 19.3 Berkshire House: construction updates

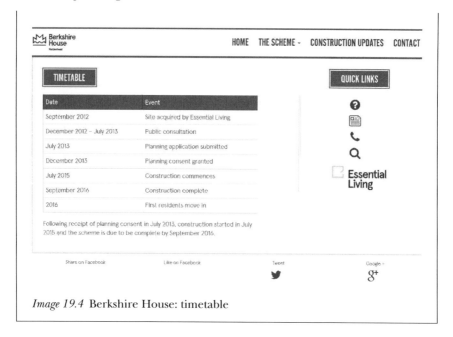

Image 19.4 Berkshire House: timetable

- A simple means of sharing news about the development is to provide plastic windows in hoardings, enabling local residents to view progress on site. This can also be provided through the use of a webcam or a series of photographs, hosted on a website or social media page.
- Other ideas used to encourage local residents to engage with the development team include the creation of community reporters (local people given the opportunity to interview the development team and report back to the community in the form of a newspaper or blog) and a regular drop-in café to encourage direct communication between the construction team and community.
- The local media can be a useful means of providing updates to the wider community and also establishing a positive relationship with a local journalist, which can be useful in the case of complaints.
- The development team also has the opportunity to involve the community in events, such as 'topping out' a significant building or opening a play area.

External bodies to assess performance

Several local and national bodies exist to evaluate construction companies' aptitude in managing community relations and to promote best practice throughout the industry.

Box 19.15 External bodies to assess performance

The Considerate Constructors Scheme

Local and national bodies exist to evaluate construction companies' aptitude in managing community relations and promote best practice throughout the industry. Of these, the Considerate Constructors Scheme is the most widely used.

The Considerate Constructors Scheme is a non-profit-making, independent organisation founded in 1997 by the construction industry to improve its image.

The Scheme is concerned about any area of construction activity that may have a direct or indirect impact on the image of the industry as a whole. The main areas of concern fall into three categories: the general public, the workforce and the environment.

Construction sites, companies and suppliers voluntarily register with the Scheme and agree to abide by the Code of Considerate Practice. The code commits those registered with the Scheme to care about appearance, respect the community, protect the environment, secure everyone's safety and value their workforce.

Around 8,000 sites, companies and suppliers register with the Scheme every year. There are a large and growing number of construction companies and clients (including numerous local authorities and housing associations) automatically registering all their work with the Scheme as company policy.

All registered sites, companies and suppliers are monitored by the Scheme, and each displays posters around the construction site, promoting their registration. If passers-by wish to comment, the name and telephone number of the site manager or company contact is clearly displayed, alongside the freephone number of the Scheme's administration office, which has a dedicated public liaison team. Registered companies and suppliers also display a vehicle sticker or magnet, showing their unique registration number, on every vehicle used on the public highway.

The Code of Considerate Practice outlines the Scheme's expectations and describes those areas that are considered fundamental for registration:

Care about Appearance: constructors should ensure sites appear professional and well managed

  • Ensuring that the external appearance of sites enhances the image of the industry

- Being organised, clean and tidy
- Enhancing the appearance of facilities, stored materials, vehicles and plant
- Raising the image of the workforce by their appearance.

Respect the Community: constructors should give utmost consideration to their impact on neighbours and the public

- Informing, respecting and showing courtesy to those affected by the work
- Minimising the impact of deliveries, parking and work on the public highway
- Contributing to and supporting the local community and economy
- Working to create a positive and enduring impression, and promoting the code.

Protect the Environment: constructors should protect and enhance the environment

- Identifying, managing and promoting environmental issues
- Seeking sustainable solutions and minimising waste, the carbon footprint and resources
- Minimising the impact of vibration and air, light and noise pollution
- Protecting the ecology, landscape, wildlife, vegetation and water courses.

Secure Everyone's Safety: constructors should attain the highest levels of safety performance

- Having systems that care for the safety of the public, visitors and the workforce
- Minimising security risks to neighbours
- Having initiatives for continuous safety improvement
- Embedding attitudes and behaviours that enhance safety performance.

Value Their Workforce: constructors should provide a supportive and caring working environment

- Providing a workplace where everyone is respected, treated fairly, encouraged and supported
- Identifying personal development needs and promoting training

- Caring for the health and well-being of the workforce
- Providing and maintaining high standards of welfare.

The Scheme provides the Best Practice Hub, a free online resource for the industry to share their considerate practice by showcasing examples of best practice in the areas of community engagement, appearance, environment, safety and workforce.

In early 2016, the Considerate Constructors Scheme developed Building Social Value (in consultation with a number of construction companies and a local authority). Building Social Value provides construction organisations and clients with a simple way to capture the social value created through construction.

Box 19.16 External bodies to assess performance

CEEQUAL

CEEQUAL is an international evidence-based sustainability assessment, rating and awards scheme for civil engineering, infrastructure, landscaping and works in public spaces. The organisation encourages and promotes the attainment of high economic, environmental and social performance in all forms of civil engineering through identifying and applying best practice. It assists clients, designers and contractors to deliver improved sustainability performance and strategy in a project or contract. The scheme rewards project and contract teams who go beyond the legal, environmental and social minima to achieve distinctive environmental and social performance in their work.

Within its People & Communities section, a wide range of positive and adverse impacts is assessed, on people affected by the project and/or on the wider communities served by or affected by the project. It covers minimising operation- and construction-related nuisances, legal requirements, community consultation, community relations programmes and their effectiveness, engagement with relevant local groups, and human environment, aesthetics and employment.

CEEQUAL became part of the BRE Group in November 2015. The next version of the scheme is now being planned to combine the strengths of CEEQUAL and BREEAM and is expected to launch in 2017/18.

Box 19.17 Using the Considerate Constructors Scheme

Comment by Alan Smith, director of group corporate responsibility, Kier Group plc

As one of the UK's largest property, residential, construction and services firms, we have a responsibility to respect and protect the people and places we serve.

Being a considerate constructor means being a good neighbour. This includes taking steps to minimise on-site noise and vibrations, informing local people about our work, caring for the environment and keeping people safe.

Since becoming an associate member of the Considerate Constructors Scheme in 2008, we have used this national initiative to measure how our operational projects are performing, for example, through waste and energy efficiency, engaging with communities and delivering high-quality working conditions for employees and subcontractors alike.

The scheme provides a score assessed on five criteria, with a maximum of 10 points available in each. Our target is to score an average of at least 40/50 by 2020, which is well beyond the 35/50 that defines best practice.

In the 2016 awards, Kier received 48 national awards, an increase from 30 in 2015 and 28 in 2014. But as a growing business, with an increasing range of service offerings both nationally and internationally, we realise that we also need to define how we should operate wherever we are working, and whatever we are doing. We are working to create a group code of conduct for positive citizenship, which will be the minimum expected of any Kier employee or of our suppliers and subcontractors anywhere in the world.

Evaluation

While evaluation need not be reported on formally, as it is with a planning application, it has many benefits. As mentioned earlier, the development team has a responsibility to deliver in line with the construction management plan, and failure to do so can lead to withdrawal of planning consent. Therefore, the team should be aware that complaints may arise and in such circumstances it will need to demonstrate that complaints were handled correctly.

Feedback can be encouraged via feedback forms and informal dialogue. Evaluation can be both qualitative and quantitative, detailing the manner and frequency with which the team communicated with local residents, responses received and changes made subsequently. This can then be communicated positively to the local community, in the form of a 'You said, we did' document which benefits the cycle of good community relations, and used to deliver future best practice.

Evaluation also benefits best practice: using examples from schemes where community relations has been successful can inform future developments.

Conclusion

Community relations is a vital component of development and one which should flow naturally from a well-run planning consultation.

Successful community relations requires a strategic and principled approach, early engagement and a realistic and appropriate set of tactics. Whether communicating directly with residents to mitigate future problems or putting in place positive programmes of engagement, a wide variety of tactics is available. Stakeholder analysis and research should inform the choice of tactics, ensuring suitability for the community and taking into account its needs. Above all, community relations should be a visible sign that the development team wishes to engage positively.

Notes

1 Chapter 14 describes the function and method of both in greater detail.
2 See BritishLand.com/policies for an example in practice.

20 Community involvement following construction

Introduction

When the last construction vehicle has left the site, what is the developer's responsibility to the new development? Of course this varies considerably: in many commercial, residential or retail schemes the developer may continue to manage the asset, whether directly or indirectly. In other circumstances, such as the construction of highways or railways, management will be formally handed over and the developer's involvement will cease.

This chapter concerns the 'grey area' in which the developer has no formal obligation to continue to invest in the new development, but there are benefits in doing so.

There are many reasons for community involvement. The 2012 Social Value Act and 2006 Companies Act require that public and private companies respectively have a legal obligation to carry out and report on social, environmental and other impacts. But there are also commercial advantages: if they are good, stakeholder relationships can benefit both the marketability of a new development and also sales and public relations.

Creating a sustainable community

In 1998, a construction industry task force under the chairmanship of industrialist Sir John Egan was put in place to 'rethink construction'. *The Egan Review: Skills for Sustainable Communities*[1] which followed sought to clarify the term 'sustainable community', to identify those responsible for leading the delivery of sustainable communities and to recognise the skills necessary to achieve sustainable communities. The principles were well received at the time and remain extremely relevant today.

The review begins by stating,

> Sustainable communities meet the diverse needs of existing and future residents, their children and other users, contribute to a high quality of life and provide opportunity and choice. They achieve this in

ways that make effective use of natural resources, enhance the environment, promote social cohesion and inclusion and strengthen economic prosperity.

It then goes on to describe the ways in which sustainable communities are best delivered, which are summarised in Box 20.1.

Box 20.1 *The Egan Review: Skills for Sustainable Communities*: **components and sub-components of sustainable communities**

A common sub-component across all components is:

- All provision and/or activity to be high quality, well designed and maintained, safe, accessible, adaptable, and environmentally and cost-effectively provided

Social and Cultural – Vibrant, harmonious and inclusive communities:

- A sense of community identity and belonging
- Tolerance, respect and engagement with people from different cultures, background and beliefs
- Friendly, cooperative and helpful behaviour in neighbourhoods
- Opportunities for cultural, leisure, community, sport and other activities
- Low levels of crime and anti-social behaviour with visible, effective and community-friendly policing
- All people are socially included and have similar life opportunities

Governance – Effective and inclusive participation, representation and leadership:

- Strategic, visionary, representative, accountable governance systems that enable inclusive, active and effective participation by individuals and organisations
- Strong, informed and effective leadership and partnerships that lead by example (such as government, business, community)
- Strong, inclusive, community and voluntary sector (such as residents' associations, neighbourhood watch)
- A sense of civic values, responsibility and pride
- Continuous improvement through effective delivery, monitoring and feedback at all levels

---

Environmental – Providing places for people to live in an environmentally friendly way:

- Efficient use of resources now and in the future in the built environment and service provision (such as energy efficiency, land, water resources, flood defence, waste minimisation etc.)
- Living in a way that minimises the negative environmental impact and enhances the positive impact (such as recycling, walking, cycling)
- Protecting and improving natural resources and biodiversity (such as air quality, noise, water quality)
- Having due regard for the needs of future generations in current decisions and actions

Housing and The Built Environment – A quality built and natural environment:

- Creating a sense of place (such as a place with a positive 'feeling' for people, and local distinctiveness)
- Well-maintained, local, user-friendly public and green spaces with facilities for everyone including children and older people
- Sufficient range, diversity and affordability of housing within a balanced housing market
- A high-quality, well-designed built environment of appropriate size, scale, density, design and layout that complements the distinctive local character of the community
- High-quality, mixed-use, durable, flexible and adaptable buildings

Transport and Connectivity – Good transport services and communication linking people to jobs, schools, health and other services:

- Transport facilities, including public transport, that help people travel within and between communities
- Facilities to encourage safe local walking and cycling
- Accessible and appropriate local parking facilities
- Widely available and effective telecommunications and internet access

Economy – A flourishing and diverse local economy:

- A wide range of jobs and training opportunities
- Sufficient land and buildings to support economic prosperity and change
- Dynamic job and business creation

- A strong business community with links into the wider economy

Services – A full range of appropriate, accessible public, private, community and voluntary services:

- Well-educated people from well-performing local schools, further and higher education and training for lifelong learning
- High-quality, local health care and social services
- Provision of range of accessible, affordable public, community, voluntary and private services (such as retail, food, commercial, utilities)
- Service providers who think and act long term and beyond their own immediate geographical and interest boundaries

Clearly, the responsibility for a sustainable community lies both with the developer – during planning and construction and immediately following its completion – and with local authorities and other partners.

It goes without saying that early and comprehensive dialogue with local stakeholders during the planning process is vital to delivering a scheme that is genuinely sustainable, and that communication at all stages should be open and transparent, welcoming input from a wide range of partners which may have a long-term interest in the scheme.

The components and sub-components described in *The Egan Review* are themselves a set of aims and objectives which might be adapted to form the basis for a sustainable community strategy, leading to the development of appropriate tactics and providing indicators by which the work may be measured. Many local authorities and government departments continue to use a sustainable development checklist based on the components of *The Egan Review* as a means of putting in place and evaluating success.

On a functional level, a sustainable community should take into account the following, all of which need to be considered at the planning stages:

- Mixed uses where appropriate, including an appropriate mix of house types and tenures
- Disabled access
- Sensitive landscaping
- Cost-effective and easy maintenance
- Features which provide a link with the location and its history
- Community spaces
- Public spaces which take into account human behaviour
- Attention to racial and religious requirements as necessary

- Art and landscaping to inspire pride and deter vandalism
- Efficiency in planning the use of energy and other resources
- Provision of footpaths and cycle routes
- Opportunities for recycling
- Connectivity with public transport

Additionally, however, there are many non-material elements to putting in place a sustainable community. These include managing (or arranging future management of) community facilities, putting in place maintenance contracts, forging links with the wider locality, creating a community which can withstand change, assisting with the creation of residents' associations and neighbourhood watch organisations and, in some cases, revisiting the new scheme at a later stage and making changes as required.

The Egan Review was greeted positively by the development community. A simple comparison with a typical 1970s/80s community and one designed latterly with these principles as its base is testament to its success. But embracing the principles of sustainable development requires an investment: developers must be prepared not only to make additional funding commitments but to invest an interest in a new community beyond its completion. In practical terms, property and landscape maintenance is often best managed by a management company. Investment of a developers' time, however, in helping to establish residents' associations and instigating a group to oversee community facilities can have a substantial benefit in the long-term success of a new community.

Box 20.2 A view on mixed-use development post Egan

Comment by Andy Lawson, projects director, Gallagher

Gallagher Estates is one of the largest strategic land companies in the UK, promoting and developing residential and mixed-use schemes in all geographical areas. We are not a house builder but play the unique role as master developer, taking forward and coordinating all aspects of the development process from initial site assembly through to master planning, planning promotion and Section 106 negotiations before delivering serviced land to the market place.

In the past 20 years, the approach to residential master planning has changed from being primarily about housebuilding to centring instead on helping to build communities. There is a growing recognition that a sustainable development requires hard infrastructure to be supplemented with a softer infrastructure in the form of community development.

Typically, a large-scale development will comprise around 25–30% public open space, and a financial contribution is made towards community amenities.

Community development strategies on any large development emerge over time as residents enter into discussions and requirements become clear. So it is vitally important to allow for flexibility.

A developer will usually put in place exit strategies which involve managing and maintaining open spaces for a few years, and then transferring the ownership and long-term management to an appropriate organisation. This might be an organisation with a democratic mandate, such as a local authority or parish council, or private management company.

It is important that developers appreciate residents' needs for information and maintain contact with growing communities through development websites, forums, email, newsletters and, increasingly, social media.

Constructive relationships with local partners, particularly local authorities which are well placed to communicate directly with specific communities, are also important throughout the build process and beyond.

The change in emphasis from housing estates to genuinely sustainable communities allows both residents and local authorities to gain a greater understanding, and help shape, emerging communities. Undoubtedly, this results in more individual, more appropriate and better functioning communities.

Community buildings

Many large-scale mixed-use schemes contain buildings specifically intended for community use. The facilities are usually built either by the developer or by the local authority using Section 106 or CIL funding. Depending on timing, residents of the new homes may be actively involved in the build process.

Timing is an important consideration: first residents on a new development would undoubtedly welcome a community centre as a place to meet neighbours, join a functioning community and learn more about both the ensuing development and the wider locality. Within a large-scale development, a community facility can also provide a perfect location for ongoing pre-planning consultation and community relations during construction. Due to funding, however, it is rarely viable to create a community facility until a substantial number of homes are sold. Furthermore, past examples have shown that the success of community facilities is determined by local residents having been involved in their instigation. A community centre built in the early stages of a development is less likely to represent the needs of the new community than one which has been created with local involvement from its conception.

Box 20.3 Case study: community involvement in the development of a community centre

Love's Farm House – St Neots

Love's Farm is a new build development on the edge of St Neots, Cambridgeshire, with 1,490 homes, a primary school, a football club and five retail properties. In 2006, land and £300,000 in Section 106 contributions were allocated for a community centre, but this was found to be insufficient to deliver a suitable building. The project gained and lost momentum a number of times, until additional funding was secured in 2012 by including a preschool facility on the same site.

The local community association, supported by Bedford Pilgrims Housing Association, was heavily involved in plans for the community centre from the start, including a failed bid to secure Lottery funding. When the full funding was eventually secured, the association formed a team that became a separate charity and now takes full responsibility for managing the centre. Local residents contributed to the design of the building and were responsible for fitting it out, writing its policies, fundraising, managing bookings, hiring staff and so on.

The centre, called Love's Farm House, opened in October 2015 and is already running at a surplus, thanks in part to the high levels of use by local residents, but also because of the high proportion of volunteer hours that go into its management and day-to-day operations. Local volunteers have been key to Love's Farm House becoming the hub of the local community, reflecting residents' interests and responding to their needs. These volunteers have striven for high standards and attention to detail – something that can be missing from community-run buildings.

Support from the local authorities still plays a crucial role. Launching a community centre proved to be a huge task for volunteers to undertake, and where volunteers felt they have the support of the council, they have done it willingly. Where that support hasn't been forthcoming, this has had an impact on volunteer morale and threatened to jeopardise the project. Giving local residents responsibility to run their own facilities as a way to reduce costs is a false economy. However, if it's done to boost engagement, nurture community cohesion and maximise the use of the facilities, then Love's Farm House shows that it can be a great success.

Frequently the establishment of a new community facility is overseen by the developer in the early stages. Quite often, a registered provider (RP) will take on the responsibility. Not only do RPs tend to have specific social commitments to the communities that they build, but where homes are available for rent they have more of a vested interest in the future success of the community than those developers building only for private sale.

The responsibility also falls to the local authority, whose community directorate or housing team may oversee the creation of the facility and perhaps continue to oversee its management. Community groups whose role it is to put in place a community facility can benefit substantially from links to the wider community, specifically the local authority, parish council members, church and youth leaders or representatives from a local school.

Developer trusts

Due to the practical difficulties of a developer remaining on site following completion of a scheme, trusts are frequently set up to benefit a community over the long term. Funding may be put in place to build and manage a venue, create landscaping or other facilities such as allotments, or for community events. Trusts are common in NSIP applications which consult upon a range of community benefits – particularly in the case of energy projects, in which case a proportion of the profits generated is sometimes paid directly to a specific community fund or project.

Box 20.4 Case study: energy companies and community funds

Sheringham Shoal – Norfolk

In late 2012, Statoil completed Sheringham Shoal Offshore Wind Farm, a full-scale commercial offshore wind investment. The wind farm comprises 88 turbines, two offshore substations, two 132-kV submarine export cables of about 22 km each as well as a 21.6-km onshore cable and new inland substation.

In addition to providing individual sponsorship for relevant projects and events in North Norfolk, Scira Offshore Energy Limited (the joint venture company which owns the Sheringham Shoal Offshore Wind Farm) worked with the Norfolk Community Foundation to establish the Sheringham Shoal Community Fund. The fund provides grants to North Norfolk community groups, including schools and NGOs which are seeking financial assistance for projects or initiatives that focus on renewable energy, marine environment and safety, sustainability, or education in these areas.

Since its establishment in 2010, the Sheringham Shoal Community fund has made grant awards totalling over £405,000 to a wide range of organisations, including grants to North Walsham Town Council to replace 109 street lights with LED lighting; Happisburgh Coast Watch to provide heating in a lookout cabin; Norfolk Rivers Trust to research the fish populations of the salt marsh surrounding Blakeney Harbour; Kickstart Norfolk to purchase and operate five electric mopeds/scooters over a 12-month period; Sheringham Coastwatch to purchase a telescope; Action for Children/Wells Children's Centre to provide swimming lessons and lifeguarding qualifications; Worstead Queen Elizabeth Village Hall to contribute to the cost of installing solar PV panels; and Norfolk Ornithologists' Association to analyse five years' worth of birdwatching data.

Both the developer and local authority should consult widely on the setting up of a financial trust, specifically its exact purpose and limitations, the way in which it may be accessed and by whom, involving local residents and their representatives in the early stages.

Box 20.5 Case study: community funds

McCarthy & Stone

In 2015, McCarthy & Stone launched a community fund with the aim of building inclusive, engaged communities in and around its retirement developments.

Studies have shown that loneliness among older people is a bigger problem than simply an emotional experience and that social isolation is harmful to health.

McCarthy & Stone found that structured community events at its developments have not only been well received by homeowners and potential purchasers, but can create a greater sense of well-being with benefits to people's health. Consequently, McCarthy & Stone introduced a community fund, which ensures that the first year of home ownership at each development can be filled with social events, from fish and chip lunches and themed movie nights to keep fit classes and arts and crafts – all with the aim of creating an established and tight-knit community for McCarthy & Stone's homeowners.

The fund totalled £100,000 in its first year, which was spent across all new developments in 2015/2016 and is due to increase year on year.

Corporate social responsibility

Frequently, a commitment to setting up a community trust or foundation is encompassed in an organisation's broader corporate social responsibility (CSR) commitments.

Corporate social responsibility is a fairly recent term used to refer to businesses' economic, social and environmental work. Some definitions suggest that CSR is used to mitigate harm caused by corporations and to enhance reputation: certainly in the case of development, environmental initiatives can address this, but this is not usually the prime driver of CSR. The Companies Act 2006 enshrines in law the concept of 'enlightened shareholder value' in place of a director's traditional common law duty of loyalty. Consequently, it has become the norm that major companies have a CSR function. The majority of blue-chip companies will have a CSR manager, strategy and funding. At the turn of the twenty-first century, it was not uncommon for a large company's philanthropic function to serve the chair's charitable interests. Increasingly, however, CSR relates directly to the function of the business and its core values.

The development industry has a unique advantage, in that each new scheme provides a very clearly defined locality and stakeholder group. Physical changes to the community brought about by development activities can provide the ideal platform for CSR related both to broader corporate objectives and to a specific locality. Thus, many developers and construction companies' CSR strategies are focused at both levels.

Typically, CSR encompasses many of the community relations initiatives described in the previous chapter, including education, employment, the arts, community buildings, charitable donations, community projects and more besides:

People and communities

- Diversity
- Employee engagement and retention
- Safety, health and well-being
- Society and community
- Training, education and apprenticeships

Box 20.6 Case study: a local charter

British Land plc

British Land has put in place a local charter which sets out how the company builds trust and supports successful, integrated local communities, connects with its neighbours and creates places that they feel engaged with.

The local commitments outlined in the charter are as follows:

1   Connect with communities so we understand local needs
2   Improve how local communities can influence decisions at our places
3   Help local people progress by supporting local jobs and training
4   Support educational initiatives for local people
5   Grow local businesses by buying their goods and services
6   Promote well-being and enjoyment
7   Offer the local community opportunities.

Box 20.7 Case study: a local charter

British Land – Regent's Place, London

At Regent's Place in London's West End, the local team has written into the contract with supplier JPC Cleaning Services that they need to promote job positions locally, reporting on progress each month. In addition, they have identified new opportunities for local firms, ranging from the waste management contract for the 13-acre campus, to ad hoc small building work. The team is also liaising with Camden Council on how they can further promote local business opportunities.

Regent's Place security partner Ultimate and its long-term community partner the West Euston Partnership supported a group of local jobseekers through a two-week training programme, developing their employability skills. JPC Cleaning and the West Euston Partnership also partnered to support another group through a British Institute of Cleaning Science training course.

Environment

* Biodiversity
* Carbon
* Waste
* Water

Box 20.8 Case study: environmental initiatives as a form of CSR

Gunsko/BNRG Renewables – Shripney, West Sussex

Gunsko was appointed by BNRG Renewables to run a consultation on a commercial-scale ground-mounted solar PV development.

The site for the proposed development was adjacent to a local nature reserve called Bersted Brooks. Gunsko contacted representatives of the nature reserve in the earliest stages of consultation and engaged with them to establish their views and address their concerns.

Although the scheme was assessed as having no detrimental effects on the nature reserve, Gunsko and BNRG Renewables were keen to promote a 'good neighbour' approach and have a net positive impact on local biodiversity. Through conversations with representatives of the nature reserve, it emerged that there had been a long-term aspiration to convert the three small ponds located on the reserve into a larger and deeper pond. The project, however, lacked funding.

BNRG Renewables have agreed to fund the project with the view to helping local residents realise their hopes for the nature reserve.

Marketplace

- Citizenship and community engagement
- Customer experience
- Labour standards and human rights
- Materials standards
- Sustainable supply chain

Governance

- Reporting and assurance
- Risk and opportunity
- Stakeholder engagement

CSR is far from being a solely charitable activity: a company can benefit in gaining a greater understanding of shareholders, users, suppliers and staff; employee motivation, learning and development; risk management or reduction; benefits to reputation; and access to capital.

Creating a CSR strategy – whether on behalf of a company or for a single development – will unify charitable/environmental/community initiatives, preventing them from being simply a random mishmash of disparate initiatives. A strategy will ensure that CSR goals are aligned with business objectives, and as business objectives change over time, they should be routinely re-examined in relation to the CSR strategy.

Box 20.9 Case study: a strategic approach to corporate social responsibility

Kier Group plc

Kier is a leading property, residential, construction and services group, which operates across a range of sectors including defence, education, housing, industrial, power, transport and utilities. As part of its vision to be a world-class, customer-focused company, Kier has embarked on an ambitious plan to grow market share, expand its range of services and, in doing so, deliver enhanced returns for shareholders, society and the environment. The strategy is led by the belief that creating shared social and environmental value will help establish Kier's reputation as a leader in its markets and enable the company to manage risks and identify opportunities which will contribute to financial performance.

Kier's strategy for sustainable business, *Responsible Business, Positive Outcomes*, consists of 20 non-financial indicators across four categories: People and Communities, Environment, Marketplace and Governance. The strategy was put in place after extensive engagement both internally and externally through one-on-one discussions and formal stakeholder engagement focus group sessions.

Kier has a central corporate responsibility (CR) team led by the director of group corporate responsibility and a group CR steering committee. The committee comprises representatives from central functions and the four divisions, with other parties and specialists also called upon from time to time, and executive board directors to support the core team on specific issues.

Additionally, The Kier Foundation is an independently registered charity which supports Kier's charity partner and other charitable causes close to the hearts of its employees and the communities its businesses operate in. In the 2014/15 financial year, Kier donated £4.24m to local communities in hours, monies and resources. The Kier Foundation (and Kier employees) raised £330,000 for Macmillan Cancer Support against a target of £75,000 in the first 18 months of the corporate charity partnership, thereby funding 12,000 nursing hours for cancer patients.

Kier will demonstrate the success of its strategy by meeting challenging targets in each of the strategy's categories and subcategories. To deliver its target, the organisation is reliant on each of the Kier businesses engaging with social value delivery throughout the journey of any contract, from the point of tendering for a job to its long-term legacy. Methods of evaluation include:

- An employee engagement survey
- Measuring the accident incidence rate (AIR) in relation to the Health and Safety Executive industry benchmark
- Ensuring that all site supervisors hold the Construction Skills Site Supervisors Safety Training Scheme (SSSTS) certificate or a recognised equivalent qualification
- An internal means of measuring 'The Kier Effect' on social value, outlined in a social value handbook
- Use of the construction technician apprenticeship programme
- Membership of The 5% Club (requiring members to commit to achieving 5% of their workforce being apprentices, graduates or sponsored students on structured programmes within the next five years)
- Various internal and external means of measuring energy usage and levels of waste to landfill
- Associate membership of the Considerate Constructors Scheme
- The use of recognised accreditation bodies such as Constructionline, Achilles, Santia and Safety Schemes in Procurement in relation to the supply chain
- Use of the Business in the Community Corporate Responsibility Index
- An annual corporate responsibility report
- A clear structure for quarterly reports to the executive board which monitors progress against targets.

Maintaining a presence in the community

CSR initiatives can provide a means by which the development team remains attached, albeit perhaps indirectly, to a new community when the physical building of the community is complete. This has been found to have many advantages, for both the new homeowners who will undoubtedly have questions in the early stages of residence and the development team who can learn much from the development of the new community.

Box 20.10 Case study: maintaining a relationship with a community

David Lock Associates – Milton Keynes

Principals and founding partners of David Lock Associates (DLA), including its chair Will Cousins, worked for the Milton Keynes Development Corporation (MKDC) in the early stages of the city's development during the 1970s and 1980s.

MKDC no longer exists, but in its capacity as an urban design and planning consultancy, DLA remains professionally involved in the design and development of Milton Keynes. The consultancy has a strong belief that city planning and community engagement go hand in hand, so individuals have maintained a role within the community.

Will Cousins continues to be involved in a voluntary role as chair of the board of trustees of the MK Gallery. In addition to holding contemporary arts exhibitions, the gallery provides a venue for discussion and debate on the future planning of the city. It hosts groups and individuals and brings them and the wider community together to debate the issues of the day.

The gallery is ideal neutral territory for debate on the future of the city as well as being an important focal point of the community. The space it provides for use by local residents, groups and schools has significantly strengthened community relations, and many people have participated in events at the gallery, giving them a greater sense of empowerment in their town.

Measuring and demonstrating success

With CSR becoming an increasingly important corporate function, indexes have been developed to assess success in such a way that national comparisons can be drawn and thus stakeholders gain a greater understanding of businesses.

Box 20.11 Measuring success

Business in the Community

Business in the Community (BITC) is a business-led charity which works with thousands of businesses to help tackle some of the key issues facing society. BITC provides a range of services, practical guidance and creative solutions that help businesses review, improve, measure and report on sustainable business practice.

The BITC Corporate Responsibility Index is a self-assessment tool that provides an insight into how leading companies are driving responsible business practice. The CR Index follows a systematic approach to managing, measuring and reporting on responsible business practices. It has four components:

- Corporate Strategy looks at the main corporate responsibility risks and opportunities to the business and how these are being identified and then addressed through strategy, policies and responsibilities held at a senior level in the company.
- Integration is about how companies organise, manage and embed corporate responsibility into their operations through KPIs, performance management, effective stakeholder engagement and reporting.
- Management builds on the Integration section, looking at how companies are managing their risks and opportunities in the areas of community, environment, marketplace and workplace.
- Performance and Impact asks companies to report performance in a range of social and environmental impacts areas.

Assessment takes the form of an online survey, and companies follow a self-assessment process intended to help them identify both the strengths in their management and performance and gaps where future progress can be made. All submissions must be signed off at the main board level to ensure director-level commitment to the veracity of the responses to the survey.

Conclusion

Developers' and local authorities' roles continuing over the lifetime of a project will vary significantly. In part, this will be dictated the scale, nature and ownership of the scheme. The impact of a truly sustainable development will be determined by the various stakeholders' (developers, local authorities and local residents) inclination to play a part in the delivery of a sustainable community.

As with consultation at the planning stage and community relations during construction, the most successful community engagement post-construction is strategic, rather than reactive: that which is informed by research and dialogue, is based on clear aims and objectives, is well planned and managed using appropriate tactics, and can be monitored and evaluated to demonstrate success and ensure future improvements.

And, as with consultation, communication should be open and transparent, responsive, engaging, accessible and realistic. Above all, a well-executed strategy, which involves the local community both as advisors and beneficiaries, provides a positive conclusion to a development which may have been many years in the making, and can speak volumes as to a constructive partnership between the development team, public bodies and the community.

Note

1 Egan, J. (2004) *The Egan Review: Skills for Sustainable Communities.* London: Office of the Deputy Prime Minister: RIBA Enterprises.

21 Conclusion

Mr Prosser said, 'You were quite entitled to make any suggestions or protests at the appropriate time, you know.'

'Appropriate time?' hooted Arthur. 'Appropriate time? The first I knew about it was when a workman arrived at my home yesterday. I asked him if he'd come to clean the windows and he said no he'd come to demolish the house . . .'

'But, Mr Dent, the plans have been available in the local planning office for the last nine months.'

'Oh yes, well as soon as I heard I went straight round to see them, yesterday afternoon. You hadn't exactly gone out of your way to call attention to them, had you? I mean like actually telling anybody or anything.'

'But the plans were on display . . .'

'On display? I eventually had to go down to the cellar to find them.'

'That's the display department.'

'With a torch.'

'Ah, well the lights had probably gone.'

'So had the stairs.'

'But look, you found the notice, didn't you?'

'Yes,' said Arthur, 'yes I did. It was on display in the bottom of a locked filing cabinet stuck in a disused lavatory with a sign on the door saying Beware of the Leopard.'[1]

Much has changed in community engagement in planning since Douglas Adams wrote his – not only fictional and but also fantastical – *Hitchhiker's Guide to the Galaxy* in the late 1970s. And few would disagree that community engagement has changed for the better.

The political impetus to increase local involvement in planning decisions is perhaps the main force for change, and the advent of Localism in 2011 has led to a wider variety of means by which developers, neighbourhood groups and local authorities may do so.

Had it not been for these political changes, however, participation in planning and development would have increased, just as it has done in other walks of life.

Today, we choose and form our own communities, be they centred on our location or our interests, online or offline. The internet has enabled us to communicate more effectively whether within our specific communities or globally. Simultaneously, organisations – both global businesses and local authorities – understand today's requirement for openness and transparency in the decision-making process, and most have sophisticated means in place to manage the process.

We have a huge toolbox of communications tactics at our disposal. In addition to the tried and tested means of consultation, such as exhibitions, surveys, newsletters and focus groups, new forms of participatory planning enable us to involve people very effectively in the early stages of urban design, replacing largely quantitative techniques with those which enable genuine dialogue and development of ideas.

Online communication has had an enormous impact on community engagement. The potential to communicate complex information on a new development as quickly, widely and efficiently as we can today would not have been imagined at the commencement of the twenty-first century, but today it is a common expectation. Online, the tactics at our disposal are more plentiful than those offline, and innovations, such as the use of gaming and virtual reality, appeal to people's sense of fun as well as providing an important role in bringing an as-yet-unbuilt scheme to life.

Of course, online consultation is still in its infancy and will remain secondary to offline consultation until we achieve universal broadband and an end to the digital divide. Excellent monitoring is required to ensure that the messages don't spiral out of control, conversations don't become lost and knowledge gaps don't appear. Consultation fatigue cannot be addressed by simply making it easier for people to state their opinion if the methods of consultation fail to allow proper consideration of the facts. And an organisation should always be prepared to show its human face in addition to its online persona – it should not be forgotten that technology has the power to disconnect, as well as to connect. Above all, online consultation must be planned: the strategic framework outlined in Chapter 14 should integrate both online and offline tactics.

Online consultation will mature over the coming years and become better integrated with other forms of consultation. Tactics, both online and offline, are likely to become increasingly qualitative in nature, although the easy accessibility of polling websites and apps will allow individuals to generate quantitative data and use it to force change.

As protestors capitalise upon the plethora of new tools, those responsible for community involvement must be equipped to rise to the ongoing challenge: using a strategic approach, targeting, segmenting and understanding audiences, tailoring tactics to the specific need and enabling results to be effectively analysed and evaluated.

Were Arthur's house to be compulsorily purchased and demolished to make way for HS2, 3, or 4 or Crossrail 3, 4, or 5, we would expect his experience to be a far more positive one.

Note

1 Adams, D. (1979) *The Hitchhiker's Guide to the Galaxy*. London: Palgrave Macmillan.

Appendix 1
Timeline of political events impacting on consultation

1835   Municipal Corporations Act establishes directly elected corporate boroughs in place of self-electing corporations which had become widely discredited and often corrupt

1888   Local Government Act establishes 62 elected county councils, including the London County Council and 61 all-purpose county borough councils in England and Wales

1894   Local Government Act revives parish councils and establishes 535 urban district councils, 472 rural district councils and 270 non-county borough councils

1899   London Government Act sets up 28 metropolitan borough councils in London and the Corporation of London

1919   Housing and Town Planning Act

1923   Housing Act

1925   Royal Commission on Local Government reports for the first time, under the chairmanship of the Earl of Onslow, with a further report published in 1929

1929   Local Government Act

1930   Housing Act

1932   Town and Country Planning Act

1935   The Restriction of Ribbon Development Act, designed to prevent the sprawl of towns and cities across the countryside

1947   Town and Country Planning Act encourages local authorities to include green belt proposals in their development plans

1953   Town and Country Planning Act

1954   Town and Country Planning Act

1955   The national green belt system is put in place to prevent urban sprawl

1959   Town and Country Planning Act

1963   London Government Act creates 32 London boroughs and a Greater London Council (GLC)

1966   Royal Commission on Local Government in England is established under the chairmanship of Lord Redcliffe-Maud (reports in 1969)

1966   Local Government Act

1967   Land Commission Act

1968   County structure plans are introduced to coordinate and guide local plans

1968   Town and Country Planning Act introduces structure plans

1969   Housing Act

1969   The Skeffington Report

1970   *Reform of Local Government in England* white paper supports most of the findings of the 1966 Royal Commission, particularly in respect of unitary local government

1971   *Local Government in England* white paper rejects unitary authorities and supports two-tier local government throughout England and Wales

1972   Local Government Act removes borough councils in rural and semi-rural areas, reduces the number of county councils in England and Wales to 47, establishes 6 metropolitan county councils and 36 metropolitan district councils

1973   Land Compensation Act

1974   Housing Act

1977   *The Crisis in Planning* – a report by the Town and Country Planning Association

1980   Local Government Planning and Land Act establishes compulsory competitive tendering and urban development corporations

1983   *Streamlining the Cities* white paper proposes to abolish the Greater London Council and six metropolitan county councils and replace them with joint boards and ad hoc agencies

1985   Local Government (Access to Information) Act

1985   The Conservative Government's white paper *Lifting the Burden* is published

1986   Committee of Inquiry into the Conduct of Local Authority Business chaired by David Widdecombe reports on political organisation of local government

1986   Housing and Planning Act

1988   Local Government Finance Act replaces domestic rates with the Community Charge or 'poll tax'

1988   Regional planning guidance is introduced as a strategic guide for county structure plans

1989   Local Government and Housing Act implements the Widdecombe report on political organisation and establishes basis of present capital finance system

1990   The Town and Country Planning Act divides planning into forward planning and development control and puts in place Section 106 agreements

1991   The Planning and Compensation Act introduces the plan-led system, affirming that planning applications should be decided in line with a development plan

1992   Local Government Finance Act replaces the Community Charge with Council Tax

1992   Local Government Act supports further structural reorganisation to create new unitary councils

1995   Publication of The Nolan Principles

1996   Select Committee on Relations Between Central and Local Government chaired by Lord Hunt of Tanworth, publishes its report, *Rebuilding Trust*, supporting change and experimentation in political organisation

1997   Local Government and Rating Act supports new opportunities for parish council formation

1997   Local Government Finance (Supplementary Credit Approvals) Act provides a basis for release of capital receipts

1997   Local Government (Contracts) Act confirms authorities' powers to enter into contracts

1998   *A Mayor and Assembly for London* white paper proposes the establishment of a Greater London Assembly with a separate directly elected mayor for London, and subsequently receives overwhelming support in a referendum

1998   *Modern Local Government: In Touch with the People* white paper

1998   Aarhus Convention

1998   Regional Development Agencies Act

1999   The Local Government Act includes a requirement for all local authorities to consult with taxpayers, and statutory guidance emphasises the need to consult widely while undertaking Best Value reviews

2000   The Local Government Act requires local authorities to prepare community strategies to promote economic, social and environmental well-being and bring about sustainable development

2000   Freedom of Information Act

2001   Publication of a white paper *Strong Local Leadership – Quality Public Services*, states that, 'We will support councils in their efforts to lead their communities and meet people's needs. In particular we will support greater levels of community engagement and involvement in council business.'

2001   *Strong Local Leadership – Quality Public Services* white paper

2004   *Community Involvement in Planning* white paper explains the government's commitment to public involvement in advance of the 2004 Planning and Compulsory Purchase Act

2004   *Releasing Resources from the Front Line*, an independent review of public sector efficiency, is published by Sir Peter Gershon, largely concerning the debate about public services and the role of government

2004   Planning and Compulsory Purchase Act

2005   *Citizen Engagement and Public Services: Why Neighbourhoods Matter*, published by the ODPM states, 'Government departments have adopted a common framework for building community capacity and agreed a shared objective: to increase voluntary and community engagement, especially amongst those at risk of social exclusion, and increasing the voluntary and community sector's contribution to delivering public services.'

2006   *Strong & Prosperous Communities* white paper established a further change in direction for local government, with a heavy emphasis on devolving power and increasing public participation and engagement

2008   *Prosperous Places: Taking forward the Review of Sub-National Economic Development and Regeneration* introduces Regional Strategies which would combine the objectives of the Regional Spatial Strategy and Regional Economic Strategy introduced by the 2004 Act

2008   Planning Act

2008   Green paper *A Stronger Society: Voluntary Action for the 21st Century*

2009   Local Democracy, Economic Development and Construction Act creates greater opportunities for community and individual involvement in local decision-making

2010   *Building the Big Society* policy paper

2010   Community Infrastructure Levy Regulations

2011   Localism Act

2011   *Open Public Services* white paper

2012   Regional development agencies abolished

2012   National Planning Policy Framework

2013   Devolution deals

2013   Growth and Infrastructure Act
2013   Revised change of use consent (an amendment to The Town and Country
        Planning [General Permitted Development] [England] Order 2013)
        first implemented the case of commercial to residential use; later
        extended to include additional use types
2015   Infrastructure Act
2016   Housing and Planning Act

Appendix 2

Examples of material and non-material planning considerations

Planning Aid England and the RTPI[1] have provided a comprehensive list of material and non-material planning considerations to guide planning consultants, as follows:

Material Planning Considerations:

- Local, strategic, national planning policies and policies in the local plan
- Emerging new plans which have already been through at least one stage of public consultation
- Pre-application planning consultation carried out by, or on behalf of, the applicant
- Government and Planning Inspectorate requirements – circulars, orders, statutory instruments, guidance and advice
- Previous appeal decisions and planning inquiry reports
- Principles of Case Law held through the courts
- Loss of sunlight (based on Building Research Establishment guidance)
- Overshadowing/loss of outlook to the detriment of residential amenity (though not loss of view as such)
- Overlooking and loss of privacy
- Highway issues: traffic generation, vehicular access, highway safety
- Noise or disturbance resulting from use, including proposed hours of operation
- Smells and fumes
- Capacity of physical infrastructure, e.g. in the public drainage or water systems
- Deficiencies in social facilities, e.g. spaces in schools
- Storage and handling of hazardous materials and development of contaminated land
- Loss of or effect on trees
- Adverse impact on nature conservation interests and biodiversity opportunities
- Effect on listed buildings and conservation areas
- Incompatible or unacceptable uses
- Local financial considerations offered as a contribution or grant

- Layout and density of building design, visual appearance and finishing materials
- Inadequate or inappropriate landscaping or means of enclosure

Non-Material Planning Considerations

- Matters controlled under building regulations or other non-planning legislation e.g. structural stability, drainage details, fire precautions, matters covered by licences etc.
- Private issues between neighbours e.g. land/boundary disputes, damage to property, private rights of access, covenants, ancient and other rights to light etc.
- Problems arising from the construction period of any works, e.g. noise, dust, construction vehicles, hours of working (covered by Control of Pollution Acts).
- Opposition to the principle of development when this has been settled by an outline planning permission or appeal
- Applicant's personal circumstances (unless exceptionally and clearly relevant, e.g. provision of facilities for someone with a physical disability)
- Previously made objections/representations regarding another site or application
- Factual misrepresentation of the proposal
- Opposition to business competition
- Loss of property value
- Loss of view

Note

1 Planning Aid England and the RTPI. The Planning Pack – Sheet 7: Development Management: Consultation and Commenting on Planning Applications (2012). London: Planning Aid England and the RTPI.

Appendix 3

Community involvement strategy outline

1   Research

    1.1   Pre-consultation dialogue

        • Initial discussions with key stakeholders

    1.2   Situational analysis

        • PEST (political, economic, social and technological) analysis
        • SWOT (strengths, weaknesses, opportunities, threats) analysis

    1.3   Stakeholder/publics research and analysis

        • Stakeholder mapping (community audit)
        • Political audit
        • Office for National Statistics/Acorn consumer classification
        • Demographic change – past and projected
        • Town/parish council information and meeting records
        • Neighbourhood forum information and meeting records
        • Relevant planning documents and comments received by the local authority on previous applications in the vicinity
        • Minutes of planning committee meetings
        • Discussions with neighbours and 'key opinion formers' where possible
        • Local press coverage and intelligence from journalists
        • Local blogs and community websites
        • Social media
        • Information about local groups (usually available from the local authority's community liaison officer)

    1.4   Issues analysis

        • An examination of the topics unearthed through research and the relevance to the issues in the proposals.

2   Plan

    2.1   Aims and objectives

2.2   Messages

2.3   Strategic overview – who, what, how and when

- Legal requirements
- Target audience
- Subject of consultation
- Extent of consultation
- Timings
- Role of anonymity in responses
- Anticipated level of participation
- Decision-making process

2.4   Consultation mandate to state:

- The organisation running the consultation
- The target audience
- The aims and objectives of the consultation
- The subject for discussion
- Potential impact of consultation
- The organisation initiating the change post-consultation

2.5   Resource allocation

- Timing
- Human resources
- Financial resources
- Consultation identity

2.6   Selection of tactics – consider:

- Accessibility
- Anonymity
- Appeal
- Balance innovation and more established methods
- Consider past successes
- Cost
- Ease
- Hard-to-reach groups
- Means of analysis
- Mix old and new methods
- Time
- Variety

2.7   Tactics timetable

2.8   Determine monitoring

3   Involve

3.1   Implement timetable

3.2   Implement monitoring

4   Evaluation and analysis

    4.1. Analysis of the results as determined in 2.6
    4.2 Evaluation – taking into account:

- Preparation
- Stakeholder identification
- Publicity
- Methods
- Responses
- Outcomes
- Feedback

5   Feedback

    5.1 Reporting and feedback – to include

- Consultation report (Statement of Community Involvement) outlining tactics deployed, results and systematic responses to feedback received
- Direct contact (usually by email) with those who responded to the consultation
- Press release

Appendix 4

Sample content for consultation websites user guides

[Scheme]

User guide and terms and conditions

1 Welcome

Thanks for taking part in the consultation on [Scheme]. We are keen to hear the views of everybody interested in the proposals and hope that you'll find this website a straightforward and enjoyable way to share your thoughts.

2 Taking part in the consultation

2.1 The process of the consultation

Our public consultation will run over an approximate number month period leading up the submission of our planning application. During this time, we will share our master plan and some more specific ideas for the development with you.

Using the [Scheme] website, you can provide feedback through polls, forums, blogs, uploading images and commenting on our updates. You can also register for updates to the website, so that you're always alerted to new content.

Additionally, we'll be running a public exhibition and meeting with local groups.

At the end of our consultation, all of the feedback received will be summarised and submitted to [Council name] District Council ('the Council') in the form of a Statement of Community Involvement, detailing the process of the consultation and an analysis of the results. Any changes to the application scheme which result from the feedback will be identified.

The Council will then run a separate consultation prior to determining whether to resolve to grant planning permission at a meeting of its planning committee. Local residents, businesses or any other third party will have the opportunity to submit representations directly to the Council prior to the meeting of the planning committee. Such representations should be sent to the Council and not to us.

We hope to receive a resolution to grant planning permission by [date].

2.2 *The importance in registering*

We ask you to register because it is important to us to know where you live when we analyse the results.

The results will be passed on to the local authority, but we undertake never to pass on your details to any other party without your express permission.

2.3 *How to contribute to polls and forums*

Before you take part in the polls, forums etc., you'll need to register. You can do this by clicking on the link on the page of the specific poll or forum, or by clicking on the **Log In** box on the top right hand corner of the web-page. The word **Register** appears at the bottom.

The registration process requires a username, password, email address, name and postcode. You'll also be given an opportunity to receive an email each time the website is updated.

You'll then be asked to activate your username by clicking on a link. The link takes you back to the website, from where you can log in using the **Log In** box on the top right hand corner of the website. Tick the '**Remember me**' if you would like your computer to remember your username and password. This is not advised if you are using a shared computer – such as in a place of work or internet café.

If you need help in taking part in the consultation, please email [dedicated email].

2.4 *Deletion policy*

Our intention is to allow respondents to comment freely. We will not delete a comment unless it is deemed to be offensive, harassment or spam or threatens the aims of the consultation (for example, if excessive comments prevent others from putting forward their point of view).

2.5 *Quick links*

On the right hand side of each page, you'll see a list of shortcut keys. Hover your cursor over these keys and they'll show you what they do. You can use these keys to access other pages on the website which are relevant to the page that you're on.

Using the links at the bottom, you can also **Share on Facebook**, **Like on Facebook**, **Tweet** and share on **Google+**.

2.6 *Signing up to alerts*

If you'd like to receive an email each time the site is updated, you can do so on registering by ticking the box '**Would you like to receive an email when this website is updated**?'

Alternatively, select the **Contact** tab at the top of the page and click on **Register for updates**.

2.7 How the results will be analysed and used

At the end of our consultation, we will analyse the results and comments and will consider whether the proposals should be adapted in the light of the comments received.

A Statement of Community Involvement will document this process and is an important element of our planning application. This will be available for you to view on the [Scheme] website, with the principal comments anonymised.

3 Terms and conditions

3.1 Tracking users

We track how you use the [Scheme] website, but unless you register we don't collect or store your personal information (e.g. your name or address) while you're browsing. This means that you can't be personally identified.

3.2 Accessibility

We're constantly working to make the [Scheme] website as accessible and usable as possible.

You can find useful guidance about the following common accessibility questions by clicking on the following external links:

- making your mouse easier to use
- using your keyboard to control your mouse
- alternatives to a keyboard and mouse
- increasing the size of the text in your web browser
- changing text and background colours
- how to magnify your screen
- screen readers and talking browsers

The website doesn't have a separate accessibility statement. This is because we've tried to design the [Scheme] website to be as accessible and usable as possible for every user.

3.3 Privacy policy

We collect certain information or data about you when you register.
This includes:

- questions, queries or comment that you leave, including your email address if you send an email to us

- your IP address, and details of which version of web browser you used
- your name, address, email address, and any other personal information that you provide

This helps us to:

- improve the site by monitoring how you use it
- respond to any feedback you send us, if you've asked us to
- provide you with information if you want it
- create a Statement of Community Involvement to accompany the planning application.

Unless you register, we can't personally identify you using your data.

3.4 Data storage and safety

We store your data on our secure servers in the UK. By submitting your personal data, you agree to this.

Transmitting information over the internet is generally not completely secure, and we can't guarantee the security of your data. Any data you transmit is at your own risk.

We have procedures and security features in place to try and keep your data secure once we receive it.

We won't share your information with any other organisations for marketing, market research or commercial purposes, and we don't pass on your details to other websites.

3.5 Content and legal rights

The [Scheme] is maintained for your personal use.

You agree to use this site only for lawful purposes, and in a manner that does not infringe the rights of, or restrict or inhibit the use and enjoyment of, this site by any third party.

We aim to update our site regularly, and may change the content at any time.

By using this site, you indicate that you accept these terms of use and that you agree to abide by them. If you do not agree to these terms of use, please refrain from using the site.

[Company name], the provider of this site, is registered with the Information Commissioner's Office under registration reference [reference]. A copy of the registration certificate is available on request.

3.6 Linking to and from the [Scheme] website

We welcome and encourage other websites to link to the information that is hosted on these pages, and you don't have to ask permission to link to the [Scheme] website.

However, we don't give you permission to suggest that your website is associated with or endorsed by the [Scheme].

Where our site contains links to other sites and resources provided by third parties, these links are provided for your information only. We have no control over the contents of those sites or resources, and accept no responsibility for them or for any loss or damage that may arise from your use of them.

3.7 Using [Scheme] content

You can reproduce information from the [Scheme] website.

We make much of our information available through feeds to third parties for use on websites or other applications. Please be aware, however, that these are not our products. These applications may use versions of our information which have been edited or cached. We don't provide any guarantees, conditions or warranties as to the accuracy of any such third-party products and do not accept liability for loss or damage incurred by users of such third-party products under any circumstances.

3.8 Disclaimer

While we make every effort to keep the [Scheme] website up to date, we don't provide any guarantees, conditions or warranties as to the accuracy of the information on the site.

We don't accept liability for loss or damage incurred by users of the website, whether direct, indirect or consequential, whether caused by tort, breach of contract or otherwise, in connection with our site, its use, the inability to use, or results of the use of our site, any websites linked to it and any materials posted on it. This includes loss of:

- income or revenue
- business
- profits or contracts
- anticipated savings
- data
- goodwill
- tangible property
- wasted management or office time

This does not affect our liability for death or personal injury arising from our negligence, nor our liability for fraudulent misrepresentation or misrepresentation as to a fundamental matter, nor any other liability which cannot be excluded or limited under applicable law.

3.9 Virus protection

We make every effort to check and test material at all stages of production; however, you must take your own precautions to ensure that the process which you employ for accessing this website does not expose you to the risk of viruses, malicious computer code or other forms of interference which may damage your own computer system.

We can't accept any responsibility for any loss, disruption or damage to your data or your computer system which may occur whilst using material derived from this website.

You must not misuse our site by knowingly introducing viruses, trojans, worms, logic bombs or other material which is malicious or technologically harmful. You must not attempt to gain unauthorised access to our site, the server on which our site is stored or any server, computer or database connected to our site. You must not attack our site via a denial-of-service attack or a distributed denial-of-service attack.

By breaching this provision, you would commit a criminal offence under the Computer Misuse Act 1990. We will report any such breach to the relevant law enforcement authorities and we will cooperate with those authorities by disclosing your identity to them.

3.10 Governing law

These terms and conditions shall be governed by and construed in accordance with the laws of England and Wales. Any dispute arising under these terms and conditions shall be subject to the exclusive jurisdiction of the courts of England and Wales.

We accept no liability for any failure to comply with these terms and conditions where such failure is due to circumstance beyond our reasonable control.

If we waive any rights available to us under these terms and conditions on one occasion, this does not means that those rights will automatically be waived on any other occasion. If any of these terms and conditions are held to be invalid, unenforceable or illegal for any reason, the remaining terms and conditions shall nevertheless continue in full force.

3.11 Revisions to these terms

We may at any time revise these terms and conditions without notice.

4 Contact details

For further information on the consultation or to provide feedback on the process, please email [dedicated email].

[Company name]
[Date]

Glossary

Accessibility The freedom for people to take part, including elderly and disabled people, those with young children and those who may encounter discrimination.

Adopted Final agreed version of a document or strategy accepted through a formal resolution. Adoption is the point at which a planning document becomes official policy.

Affordable housing Social rented, affordable rented and intermediate housing provided to eligible households whose needs are not met by the market. Eligibility is based on local incomes and local house prices.

Alt text (alternative text) A word or phrase that can be inserted in an HTML document to tell website viewers the nature or content of an image. The alt text appears in a blank box that would normally contain the image.

Amenity A positive element that contributes to the overall character or enjoyment of an area. For example, open land, trees, and historic buildings or less tangible factors such as tranquillity. Residential amenity considerations may include privacy (overlooking), overbearing impact, overshadowing or loss of daylight/sunlight.

App (application) A type of software programme that can be downloaded onto a computer, tablet or smart phone.

Appeals The process whereby a planning applicant can challenge a decision, including refusal of planning consent. Appeals can also be made against the failure of a planning authority to issue a decision within a given time and against conditions attached to a planning permission. In England and Wales, appeals are processed by the Planning Inspectorate.

Area action plan A type of development plan document focused upon a specific location or an area subject to conservation or significant change (for example, a major regeneration scheme).

Area of Outstanding Natural Beauty (AONB) An area with statutory national landscape designation, the primary purpose of which is to conserve and enhance natural beauty.

Arnstein's Ladder of Citizen Participation   A theory put forward by Sherry Arnstein in 1969 which presented a deliberately provocative take on the relationship between community and government by using a ladder as a metaphor for increasing access to decision-making power.

Article 4 Direction   A direction removing some or all permitted development rights, for example within a conservation area or curtilage of a listed building. Article 4 directions are issued by local planning authorities.

Asset of Community Value (ACV)   Buildings or land identified under the Localism Act 2011 as being of importance to a community's social well-being, which as such afford a 'Community Right to Bid'.

Authentication   The process of verifying a user's identity prior to an online transaction, such as a domain name transfer.

Authorisation   The process of verifying that a user is authorised to perform an action, such as make changes to a domain name's contact information.

Average session time   The total duration of all sessions (in seconds) spent on a website, divided by the total number of sessions.

Blog   An abbreviation of weB LOG: a journal that is available online and is updated by the owner regularly.

BME   Black and minority ethnic: terminology used to describe people of non-White descent.

Bounce rate   The percentage of visitors to a website who navigate away from the site after viewing only one page.

Broadband   A high-speed internet connection.

Brownfield land   Land which is or was occupied by a permanent structure. See also *Previously Developed Land.*

Browser   Computer software which can be used to search for and view information on the internet.

Budget simulator   A tool, usually used by local authorities, to engage citizens in budget decisions.

Build-to-rent   One of a series of government initiatives to increase the supply of high-quality homes available for market and affordable/discounted rent in the private sector.

Bulletin board (online)   An online noticeboard

Business improvement district (BID)   Designated town centre management (and sometimes other areas) where businesses agree to pay additional rates to fund improvements to the general retail environment.

Call-in   The Secretary of State for Communities and Local Government can order that a planning application or local plan is taken out of the hands of a local authority. The application will then be subject to a public inquiry presided over by a planning inspector who will make a recommendation to the secretary of state.

Change of use   Planning permission is required for the 'material change of use' of a property except in specific circumstances.

Character A term often relating to conservation areas or listed buildings, but also to the more general appearance of a vicinity.

Charrette A workshop devoted to utilising the thoughts of all those present to solve a problem or develop a concept.

Chat room An online service that allows users to communicate with each other about an agreed-upon topic in 'real time' as opposed to delayed time with e-mail.

Choices method A technique used to model a decision process via revealed preferences or stated preferences made in a particular context. Typically, it attempts to use discrete choices (A over B; B over A, B and C) in order to infer positions of the items on some relevant latent scale.

Citizen advisory committee Committees which include citizens or community representatives to advise on policymaking or decisions.

Citizen jury/panel A group of people chosen to represent the community or communities and input into the decision-making process.

Code for Sustainable Homes A national standard for sustainable design and construction of new homes.

Code frames (in analysis) The list of values and their associated interpretations in a question.

Co-design An approach to planning which attempts to actively involve all stakeholders (e.g. employees, partners, customers, citizens, end users) in the design process to help ensure the result meets their needs. See also *Participatory design*.

Collaborative planning The process of involving the wider community in a planning project.

Community asset Land and building owned or managed by a community organisation. This can include town halls, community centres, sports facilities, affordable housing and libraries.

Community Asset Transfer The transfer of responsibility for buildings or land from a local authority to a voluntary or community organisation.

Community benefits Aspects of a proposed development which bring about social, economic or environmental benefits. Community benefits may be put in place to mitigate the impact of development.

Community engagement Actions and processes taken or undertaken to establish effective relationships with individuals or groups within a defined community so that more specific interactions can then take place.

Community facility An asset provided for the benefit of the community. This might include a community centre, church or library.

Community infrastructure levy (CIL) A planning charge, introduced by the Planning Act 2008, as a tool for local authorities in England and Wales to deliver infrastructure to support development.

Community involvement Effective interactions between applicants, local authorities, decision makers, individual and representative stakeholders to identify issues and exchange views on a continuous basis.

Community Land Trust (CLT)   A non-profit organisation that develops and stewards affordable housing, community gardens, civic buildings, commercial spaces and other community assets.

Community liaison officer   A person who liaises with the community on behalf of either a local authority or development team to enable the two organisations to communicate and work together.

Community of interest   A group of people or organisations with a shared concern who have united to campaign for a common cause

Community planning   An approach to planning which puts the community at the heart of its activities.

Community planning day   Usually occurring at the start of a planning process, community planning days enable the wider community to work with professionals to explore options for a site, neighbourhood or city. A community planning day usually involves presentations and workshops and is led by an external facilitator.

Community Right to Bid   An initiative introduced under the Localism Act to give community organisations the opportunity to purchase and manage assets that are of value to the local community.

Community Right to Build   An initiative introduced under the Localism Act to enable a small-scale development of community facilities or housing to be built to meet a local need.

Community Right to Build Order   An order made by the local planning authority (under the Town and Country Planning Act 1990) that grants planning permission for a site-specific development proposal or class of development.

Community Right to Challenge   An initiative introduced under the Localism Act which enables voluntary and community bodies and employees of the local authority who wish to form a mutual organisation to take over the delivery of a specific service from the local authority.

Community Strategy   A requirement under the Local Government Act 2000 to promote and improve the economic, social and environmental well-being of an area and to contribute to achieving sustainable development.

Community visioning   The process of a specific local community developing a plan, goal or vision for future development.

Compulsory Purchase Order (CPO)   An order issued by government or a local authority to acquire land or buildings in the wider public interest, for example, for the construction of a major road.

Concept statement   A concise, diagrammatic illustration of ideas which sets out the vision and broad principles for the development of a site and will set the context for the design and access statement.

Conditions (on a planning permission)   Requirements attached to a planning permission that limit or direct the manner in which development is carried out. Should these be breached, the local authority can take enforcement action.

Consensus building   A collaborative approach to problem solving which seeks solutions that are agreeable to all parties.

Conservation   The process of maintaining and managing change to a heritage asset in a way that sustains and, where appropriate, enhances its significance.

Conservation area   An area of special architectural or historic interest, the character, appearance or setting of which is desirable to preserve or enhance; permitted development rights may be restricted in these areas.

Conservation society   An organisation established with the prime objective of protecting and preserving the environment. Typically, conservation societies exist to protect original architecture or natural resources.

Construction impacts group   A group set up, often by a developer, to mitigate the negative impact of development on a community.

Construction management plan (CMP)   A set of conditions put in place by a local authority to ensure that developers minimise the negative impact of construction.

Consultation   The process of sharing information and promoting dialogue between local planning authorities, applicants, individuals or civic groups, with the objective of gathering views and opinions on planning policies or development proposals.

Consultation café   A local café run, usually by a local authority or developer, to further the process of consultation through dialogue and other means.

Consultation fatigue   The reluctance to take part in consultation, usually because of excessive past consultation or lack of demonstrable results from previous consultation.

Consultation hub   The central point of a consultation. This may be online (e.g. a consultation café) but is more frequently used to describe a consultation website which provides the basis for online consultation.

Consultation mandate   A document which sets out the parameters and process of the consultation.

Consultation panel   A group of people, usually representative of the wider community, who meet to comment on local plans and applications on behalf of their local community.

Consultation report   A document produced at the end of a consultation process which summarises the process of consultation, the results achieved and an analysis of those results. See also *Statement of Community Involvement.*

Core strategy   A local plan document which sets out the long-term vision strategic objectives and strategic planning policies for a local authority area.

Corporate social responsibility (CSR)   The recognition that a company or organisation should take into account the effect of its social, ethical and environmental activities on its staff and the community around it.

Decision notice   A legal document which states a local authority's decision on a planning application. The notice includes any conditions attached to the permission or in the case of a refusal, the detailed reasons for the refusal.

Delegated powers   Powers conferred to designated planning officers by locally elected councillors to enable officers to take decisions on planning matters.

Deliberative polling   A form of consultation that combines the techniques of public opinion research and public deliberation. A sample is polled on a specific issue and then some of the sample are invited to an event to discuss the issue.

Deliberative workshop   A form of facilitated group discussion that gives participants the opportunity to consider an issue in depth, challenge opinions and reach an informed position.

Department for Communities and Local Government (DCLG)   The UK government department with responsibilities for housing, planning and development.

Design and access statement   A concise report accompanying certain applications for planning permission or for listed building consent. The report provides a framework for applicants to explain how a proposed development is a suitable response to the site and its setting, and demonstrate that it can be adequately accessed by prospective users.

Design code   A set of illustrated design rules and requirements which instruct and may advise on the development of a site or area.

Design guide   A document, often produced by a local authority, providing guidance on how development can be carried out in accordance with good design practice and to retain local distinctiveness.

Design statement   A document describing the design principles upon which a planning proposal is based. The design statement may form part of a planning application.

Detailed application/full application   A planning application seeking full permission for a development proposal, with no matters reserved for later planning approval.

Developer trust   A fund put in place by a developer to provide an ongoing financial resource to the community, usually for a specific purpose.

Development brief   A document produced to provide information about preferred option(s) for the development of a site.

Development consent order   The means of obtaining permission for a development categorised as nationally significant infrastructure project (NSIP).

Development control/management   The process whereby a local planning authority determines planning applications. Local authorities must take into account relevant development plan policies, financial considerations and any other material considerations.

Development plan document (DPD) A spatial planning document prepared by the planning authority that is subject to an independent public examination. DPDs can cover a range of issues and will set out the main spatial strategy, policies and proposals of the local authority. DPDs include the core strategy, adopted proposals map, site specific allocations and area action plans.

Development trust A community organisation which aims to address a range of economic, social, environmental and cultural issues, seeking to reduce dependency on grant support by generating income through enterprise and the ownership of assets.

Digital divide The gap between those who have access to technology and those who do not, typically due to availability of technology and network coverage but also money, location and literacy.

Discussion board (online) An online 'bulletin board' where individuals can post messages and respond to others' messages.

Discussion forum (online) See *Discussion board (online)*.

Discussion group (online) See *Discussion board (online)*.

Document library A resource, often online, providing documents relevant to a planning application. This might include the planning statement, design and access statement, retail statement and transport assessment.

Domain name The unique name that identifies an internet site. Domain names are composed of two or more parts separated by dots. The first part is the most specific, and the second part is an extension relating to either a country code and/or the category of the domain, e.g., de (Germany) or. co.uk (UK company) or ac.uk (academic institution in UK) or. com.

Download The process of transferring data from a remote computer to a local computer.

Drop-in event Similar to an exhibition and often used interchangeably, but sometimes with fewer illustrative materials and with the opportunity to speak one to one with a consulting body.

Duty to cooperate A measure within the Localism Act which ensures that local planning authorities work with neighbouring authorities and other public bodies to address strategic issues that affect local plans and cross administrative boundaries.

E-government/E-planning Government initiatives to assist local authorities in providing planning services online.

Email campaign A campaign conducted through multiple or bulk email lists.

Enforcement A means by which local authorities may take action against development which has not been properly authorised.

Enforcement Notice A formal notice served by the local planning authority requiring the owner of a property to remedy a breach of planning control.

Enquiry by Design   A form of community planning which uses charrettes and other methods to involve stakeholders in design projects.

Environmental impact assessment (EIA)   A procedure to ensure that decisions are made in full knowledge of any likely significant effects on the environment.

Ethnic minority   Any ethnic group in Britain except White British.

Evidence base   The information and data gathered by local authorities to justify the 'soundness' of the policy approach set out in local development documents, including physical, economic, and social characteristics of an area.

E-voting   Electronic (online) voting.

Examination   An independent inquiry into the soundness of a draft development plan document, chaired by an inspector appointed by the secretary of state.

Extended Validation (EV)   An enhanced version of SSL (Secure Socket Layer) which utilises the same security levels as conventional SSL certificates.

FAQ   Frequently Asked Question.

Firewall   Software that helps protect a computer from viruses and intruders by creating a 'wall' between the computer and the internet.

Focus group   A small group of people whose opinions about a product or issue are explored to aid understanding on the opinions that can be expected from a larger group.

Front-loading   Community involvement in the pre-application phase of development proposals or the production of local development documents to gain public input and seek consensus from the earliest opportunity.

General Permitted Development Order (GPDO)   Regulations made by government which grant planning permission for specified minor developments.

Google Analytics   A web analytics service which tracks and reports website traffic.

Google Maps   A web mapping service which offers satellite imagery, street maps, 360° panoramic views of streets, real-time traffic conditions, and route planning for traveling by foot, car, bicycle or public transportation

Green belt   A designation for land around certain cities and large built-up areas, which aims to keep the land permanently open or undeveloped.

Greenfield   Previously undeveloped land.

Gunning principles   A set of principles, based on case law, which set out the legal expectations of consultation.

Hard-to-reach groups   Those groups of society which is particularly difficult to reach through the usual means.

Housing association   A common term for independent, not-for-profit organisations which work with local authorities to offer homes to local people.

Housing land availability (HLA)   The total amount of land reserved for residential use awaiting development.

Housing need   A level of socially desirable housing, the demand for which is not reflected in the open market and which is therefore usually met through subsidy.

HTML   Hypertext Markup Language: a standardized language of computer code, imbedded in 'source' documents behind all web documents, containing textual content, images, links to other documents and formatting instructions for display on the screen.

Hyperlink, hypertext   Text on a webpage which links to another document or webpage.

Independent examination   An examination is undertaken by an independent inspector into representations on a development plan document, including its legal compliance and overall soundness. The examination is likely to include public hearing sessions.

Informal hearing   Informal discussion of the proposal for determining planning appeals and chaired by an inspector.

Information Commissioner's Office (ICO)   An independent authority set up to uphold information rights in the public interest, promoting openness by public bodies and data privacy for individuals.

Inspector's report   A document produced by an independent inspector from the Planning Inspectorate. It assesses the soundness and robustness of planning documents.

Internet Protocol (IP) address   A unique number used to identify a computer on the internet.

Java   A network-oriented programming language specifically designed for creating programs that can be downloaded to a computer from a web page and run immediately. Using small Java programs, web pages can include features such as animations, calculators and other interactive features.

Judicial review   A court proceeding in which a judge reviews the lawfulness of a decision or action made by a public body.

Key worker living   A government scheme intended to help key workers in London, the South East and East of England to buy a home, upgrade to a family home or rent a home at an affordable price.

Lobbying   The process whereby individuals, civic groups or commercial organisations seek to influence planning decision makers by employing a variety of tactics.

Local charter   A contract between a developer and the local community, usually putting in place specific targets in relation to economic and social initiatives.

Local development document (LDD)   A local authority document constituting a planning strategy.

Local development framework (LDF)   A portfolio of local development documents which sets out the local planning authority's policies for using development to meet economic, environmental and social aims for the future.

Local development order (LDO)   The granting of planning permission by a local planning authority under the Town and Country Planning Act 1990.

Local enterprise partnership (LEP)   A voluntary partnership between a local authority and businesses set to determine local economic priorities and lead economic growth.

Local plan   The main planning policy document for a local authority area. The local plan's 'development plan' status means that it is the primary consideration in deciding planning applications.

Local planning authority   The public authority whose duty it is to carry out specific planning functions for a particular area. Includes district councils, London borough councils, unitary authorities, county councils, the Broads Authority, the National Park Authority and the Greater London Authority.

Local strategic partnership   A group of public, private, voluntary and community organisations and individuals responsible for preparing a community strategy.

Localism   The Localism Act 2011 devolved greater powers to local government and neighbourhoods and gave local communities additional rights over planning decisions.

Master plan   A document outlining the overall approach to the layout of a development.

Material consideration   Factors which must be taken into account when making a planning decision.

Message board   See *Discussion board (online)*.

Minority group   Traditionally under-represented groups, the marginalised, disadvantaged or socially excluded, e.g. Black and ethnic minority groups, gypsies and travellers or asylum seekers.

Mixed-use development   Developments constituting more than one use type.

Monitoring   Regular measurement of progress towards targets, aims and objectives. Also involves scrutiny, evaluation and, where necessary, changes in policies, plans and strategies.

Moratorium   The temporary prohibition of (development) activity.

National Infrastructure Plan   A vision, published in October 2010, outlining the future of UK economic infrastructure, setting out the challenges facing UK infrastructure and the government's strategy for meeting infrastructure needs.

National Planning Policy Framework (NPPF)   A comprehensive document which sets out the government's national planning requirements, policies and objectives. It replaces Planning Policy Statements, Planning Policy Guidance and Circulars. The NPPF is a material consideration in the preparation of LDDs and when considering planning applications.

National Planning Practice Guidance (NPPG) The bringing together of many areas of English planning guidance into a new online format to accompany the NPPF.

Nationally significant infrastructure project (NSIP) A project of a type and scale defined under the Planning Act 2008 in relation to major infrastructure developments – usually energy, transport, water and waste. These projects require a single development consent which follows a different procedure to that of standard planning applications.

Neighbourhood area A designated area within a local authority which has specific planning policies attached to it. Neighbourhood areas are created following an application made by a parish or town council or a prospective neighbourhood forum to the local planning authority under regulation 5 of the neighbourhood planning (General) Regulations 2012 (as amended).

Neighbourhood development order An order made by a local planning authority (under the Town and Country Planning Act 1990) through which parish councils and neighbourhood forums can grant planning permission for a specific development proposal or classes of development.

Neighbourhood forum A community group that is designated to take forward neighbourhood planning in areas without parishes.

Neighbourhood Plan A plan prepared by a parish or town council or a neighbourhood forum for a particular neighbourhood area. The plan can set out land use planning policies and the allocation of land, provided they are in conformity with strategic policies of the local plan and have regard to national policy and guidance. The process is subject to independent examination and a community referendum.

NIMBY An abbreviation of Not In My Back Yard used in relation to those who appear to oppose any development in the vicinity of their homes and do so for purely selfish reasons.

Objection A written representation made to a local planning authority by an individual, civic group or statutory consultee in response to local plan proposals or a planning application.

Online consultation Consultation which takes place via website, email, social media or other online means.

Online forum See *Discussion board (online)*.

Open/public meeting A meeting (usually to launch a consultation or present and discuss a development proposal) which is open to all.

Opinion poll An assessment of public opinion by questioning a representative sample, used often to test attitudes to proposed developments or emerging local plan policies.

Organic search The method of finding a website by entering search items into a search engine.

Outline application An application for planning permission which does not include full details of the proposal. Essentially an outline consent approves the principle of development, not the detail.

Pairwise analysis   A process of comparing entities in pairs to determine which entity is preferred, or has a greater amount of some quantitative property, or whether or not the two entities are identical.

Parish plan   A community planning tool which assists communities to articulate issues that concern them. This results in an action plan which can to inform and endorse a parish council's role in acting on behalf of a community.

Parish/town council   An elected local government body which provides a limited range of local public services and makes representations on behalf of the community to other organisations.

Participation   The extent and nature of activities undertaken by those who take part in community involvement.

Participatory design   An approach to planning which attempts to actively involve all stakeholders (e.g. employees, partners, customers, citizens, end users) in the design process to help ensure the result meets their needs. See also *Co-design*.

PDF   Abbreviation of Portable Document Format, a file format developed by Adobe Systems that is used to capture almost any kind of document with the formatting in the original.

Permitted development   A nationwide planning permission to carry out certain limited forms of development without the need to make a planning application. These provisions are granted under the Town and Country Planning (General Permitted Development) (England) Order 2015. Local planning authorities have the power to remove permitted development rights through planning conditions or Article 4 Directions.

PEST analysis   A framework of macro-environmental factors used in strategic management which gives an overview based on political, economic, social and technological factors.

Petition   A request to do something, most commonly addressed to a government official or public entity. Although it can take a variety of forms, a petition usually involves a list of signatures and/or names and addresses. Increasingly found online.

Phasing or phased development   The 'rolling out' of a large development into manageable parts.

Phishing (online)   The attempt to obtain sensitive information such as usernames, passwords and credit card details by disguising as a trustworthy entity.

Picture boards   A bulletin or discussion board which facilitates the posting of images.

Planning brief   Site-specific development briefs, design briefs, development frameworks and master plans that seek to positively shape future development.

Planning committee   The planning decision-making body of a local authority. The planning committee is made up of elected members. Its main role is to make decisions on planning applications.

Planning condition   A condition imposed on a grant of planning permission (in accordance with the Town and Country Planning Act 1990) or a condition included in a local development order or neighbourhood development order.

Planning for Real®   A consultation method involving a creative exercise (for example, the use of maps and model buildings) designed to engage the public in plan making.

Planning gain   Community benefit, secured by way of a planning obligation as part of a planning approval and provided at the developer's expense. For example, affordable housing, community facilities or mitigation measures. See also *Planning obligation* and *Section 106 obligations.*

Planning guidance   Non-statutory strategy and policy documents which inform local planning policies.

Planning obligation   A legally enforceable obligation entered into under Section 106 of the Town and Country Planning Act 1990 to mitigate the impacts of a development proposal. See also *Planning gain* and *Section 106 obligations.*

Political advocacy   A less emotive term for lobbying.

Political audit   A survey of the political stakeholders relevant to a specific consultation or process of decision-making.

Pop-up events   Temporary events, often held at unconventional locations.

Pre-application discussions   Meetings between a prospective applicant and the local authority prior to making a planning application, generally confidential in nature.

Preferendum   A consultation technique similar to a referendum which uses a selection of options in place of yes/no questions.

Previously developed land   Land which is or was occupied by a permanent structure. See also *Brownfield land.*

Prior approval   A procedure where permission is deemed granted if the local planning authority does not respond to the developer's application within a certain time.

Project liaison group   A group of stakeholders, usually representative of the wider community, with whom the development team discusses a development proposal on an ongoing basis.

Public art   Permanent or temporary physical works of art visible to the general public – such as sculpture, lighting effects, street furniture, paving, railings and signs.

Public inquiry   A formal procedure for dealing with planning appeals, where parties frequently have legal representation and cross-examination takes place.

Registered Social Landlord (RSL)   Organisations that provide affordable housing. Most housing associations are RSLs. They own or manage some 1.4 million affordable homes, both social rented and intermediate.

Reserved matters   An outline planning permission may specifically reserve some matters not relating to the principles of the proposed development for later consideration. These can include access, appearance, layout, scale and landscaping.

Residents' association   An association which is usually formally constituted of people living in the same building or neighbourhood which deals with matters of common interest.

Responsive web design (RWD)   An approach to web design which allows desktop webpages to be viewed in response to the size of the screen or web browser being used – e.g. a smartphone or tablet.

Retail impact assessment   An assessment undertaken for an application for retail use, exploring the impact of the proposal on existing centres within the catchment area of the proposed development.

Roadshow   A touring event. In the case of a planning consultation, this will normally refer to an exhibition which visits more than one venue.

Roundtable discussions   A forum in which people making representations upon a development plan document can express their views before a government-appointed planning inspector.

Section 106 obligations   Requirements of developers as part of planning permissions. These are agreed in the planning application process, to provide contributions (usually financial) to develop facilities/amenities for the local community (e.g. education, open space). See also *Planning gain* and *Planning obligations.*

Section 42 consultees   Under Section 42 of the Planning Act 2008, promoters of an NSIP planning application must consult statutory consultees, i.e. 'prescribed persons' listed in Schedule 1 to the Infrastructure Planning (Applications: Prescribed Forms and Procedure) Regulations 2009; local authorities prescribed in Section 43 of the Planning Act 2008; the Greater London Authority if the site is situated in Greater London; the persons prescribed in Section 44 of the Planning Act 2008 including owners, lessees, tenants, and those with an interest in the land.

Section 47 consultees   Under Section 47 of the Planning Act 2008, promoters of an NSIP planning application must consult the local community.

Silent majority   A significant proportion of a population who choose not to express their views, often because of apathy or because they do not believe their views matter.

Simplified planning zone   An area in which a local planning authority wishes to stimulate development and encourage investment. This is achieved by the granting of a specified planning permission.

Site allocation   Designation of land in a local development documents for a particular land use (e.g. housing).

Site of Special Scientific Interest (SSSIs)   Sites designated for special protection by Natural England under the Wildlife and Countryside Act 1981.

Site-specific allocations   The allocation of sites for specific or mixed uses. Policies will identify any specific requirements for the site.

Situational analysis   A collection of methods that can be used to analyse the internal and external environment in relation to a specific proposal. See also *PEST analysis* and *SWOT analysis.*

Slider bar A graphical control element on a website with which a user may set a value by moving an indicator, usually horizontally.

Smartphone A mobile phone which, as well as making calls and sending texts, can be used to browse the internet, write email and perform a number of other functions.

SMS Short Message (or Messaging) Service: a system that enables mobile phone users to send and receive text messages.

Social media Websites and applications that enable users to create and share content or to participate in social networking.

Spam Unsolicited commercial email.

Special interest group A community within a larger vicinity with a shared interest, where members cooperate to affect or to produce solutions in relation to their area of specific interest and may communicate, meet and organise events.

SSL Secure Sockets Layer: a protocol to enable encrypted, authenticated communication across the internet. SSL is used mostly, but not exclusively, in communications between web browsers and web servers. A URL that begins with 'https' instead of 'http' indicates an SSL connection will be used.

Stakeholder A person, group, company, association, etc. with an economic, professional or community interest in the proposals.

Stakeholder/publics analysis The process of analysing stakeholder groups and their (likely) views.

Stakeholder mapping The process of identifying stakeholder groups geographically.

Statement of Community Consultation (SoCC) A document used in an NSIP planning application which is prepared in accordance with Section 47 of the Planning Act 2008 and explains how the development team will consult the local community on the proposed planning application.

Statement of Community Involvement (SCI) A document which sets out the processes to be used by the local authority in involving the community in the preparation, alteration and continuing review of all local development documents and development control decisions. The term is also used to refer to the report compiled at the end of a consultation by the development team and submitted as part of the planning application.

Statutory consultees Groups with which development teams/local authorities must consult under regulations set out in national planning policy.

Sustainability appraisal (SA) A systematic and iterative appraisal process, incorporating the requirements of the European Strategic Environmental Assessment Directive. Its purpose is to appraise the social, environmental and economic effects of the strategies and policies in a local development document from the outset of the preparation process.

SWOT analysis SWOT is an acronym for strengths, weaknesses, opportunities and threats and is a structured planning method that evaluates those four elements of a project or business venture.

Think tank   A body of experts providing advice and ideas on specific political or economic problems.

Transport assessment   A comprehensive and systematic process that sets out transport issues relating to a proposed development. It identifies what measures will be required to improve accessibility and safety for all modes of travel, particularly walking, cycling and public transport and what measures will need to be taken to deal with the anticipated transport impacts of the development.

Troll   Online, a troll is a person who sows discord by posting inflammatory, extraneous or off-topic messages in a community such as a newsgroup, forum, chat room or blog.

Unique users   Unique IP addresses which have accessed a website. Calculating the number of unique users is a common way of measuring the popularity of a website.

Upload   The process of transferring files from a computer to the internet.

URL   Uniform Resource Locator: the unique address of any web document.

Use Classes Order   The Town and Country Planning (Use Classes) Order 1987 puts uses of land and buildings into various categories. Planning permission is not needed for changes of use within the same use class unless a planning condition has been imposed restricting this.

vCard   A file format standard for electronic business cards. vCards are often attached to e-mail messages and can be downloaded from websites. They can contain name and address information, telephone numbers, e-mail addresses, URLs, logos, photographs and audio clips.

Vox pop   An informal comment from a member of the public, often broadcast or published.

Web 2.0   The second stage of development of the internet, characterised especially by the change from static web pages to dynamic or user-generated content and resulting in the growth of social media.

Web mapping   The process of using maps delivered by geographical information systems (GIS). A web map on the internet can allow interaction by the user.

Webinar   An umbrella term for various types of online collaborative services including web seminars, webcasts and peer-level web meetings.

Wiki   A website that provides collaborative modification of its content and structure directly from the web browser.

Written representations   A procedure by which representations on planning appeals, development plans and development plan documents are dealt with by change of correspondence without the need for a full public inquiry or informal hearing.

Further reading

Beresford, M. Smart People Smart Places Realising Digital Local Government (2014). London: New Local Government Network (NGLN).

Bickerstaffe, S. *Building Tech-Powered Public Services* (2013). London: The Institute for Public Policy Reform (IPPR).

BSHF. Community Guide: Masterplanning (2014). https://bshf-wpengine.netdna-ssl.com/wp-content/uploads/2016/03/Community-Guide-Masterplanning_BSHF5.pdf.

Carnegie UK Trust. Click and Connect: Case Studies of Innovative Hyperlocal News Providers (2015). Dunfermline: Carnegie United Kingdom Trust.

CBI. Building Trust: Making the Public Case for Infrastructure (2014). London: CBI

Coleman, S. Connecting Democracy: Online Consultation and the Flow of Political Communication (2012). Cambridge, MA MIT Press.

Communispace. The Rules of Community Engagement (2013). Boston, MA: Communispace Corporation.

Cullingworth, B. A. *Town and Country Planning in the UK* (2006). London: Routledge.

Curtin, T. *Managing Green Issues* (2004). London: Palgrave MacMillan.

Denyer-Green, B. *Development and Planning Law* (2013). London: Routledge.

Department for Communities and Local Government. *A Plain English Guide to the Localism Act* (2011). London: Department for Communities and Local Government.

Department for Communities and Local Government. *The National Planning Policy Framework* (2014). London: Department for Communities and Local Government.

Digital Democracy Commission. *Open Up! Report of the Speaker's Commission on Digital Democracy* (2015). London: Digital Democracy Commission.

Doak, J., and Parker, G. *Key Concepts in Planning* (2012). London: Sage.

Jones, R., and Gammell, E. *The Art of Consultation* (2009). London: Biteback Publishing.

Local Government Association. Transforming Local Public Services: Using Technology and Digital Tools and Approaches (2014). London: Local Government Association.

Moore, V. *A Practical Approach to Planning Law* (2010). Oxford: Oxford University Press.

New Economics Foundation. Participation Works! 21 Techniques of Community Participation for the 21st Century (1999). London: New Economics Foundation

Planning Aid England and the RTPI. The Planning Pack – Sheet 7: Development Management: Consultation and Commenting on Planning Applications (2012). London: Planning Aid England and the RTPI.

Preece, C., Moodley, K., and Smith, A. *Corporate Communications in Construction* (1998). London: Blackwell Science.

Radcliffe, D. Here and Now: UK Hyperlocal Media Today (2012). London: Nesta.

Rydin, Y. Urban and Environmental Planning in the UK (1998). London: MacMillan Press.

Rydin, Y. *The Purpose of Planning* (2011). Bristol: Policy Press.

Wates, N. The Community Planning Handbook (2014). London: Routledge.

YES Planning, Davies, J., Gillingham, H., Godley, H., Greenwell, J., Megson, C., Turner, C., and White, R. *Planning The Future* (2014).

Index

38 Degrees 40

Aarhus Convention 16
activism 57, 278–90
Alconbury Weald 284
Alison Turnbull Associates 320
Amec Foster Wheeler 272
analysis 55, 182, *197*, 230, 243, 259–77,
    281–3, 287, 288; qualitative 205,
    226–9, 232, 243, 356; quantitative
    226, 260–1
anonymity 210–11, 231, 232
apathy 292–3
appeals 1, 31, 135, 162–76; in England
    and Wales 162–7; intervention by the
    London mayor 164; judicial review
    119, 120n17, 169–71, 175; Northern
    Ireland 168; recovery of 164;
    Scotland 167–8; statistics 168
apprenticeships 326, 347, 351
Argent (Kings Cross) Ltd 212, 223, 262,
    296, 329
Arnstein's Ladder of Participation 3
arts 38, 214, 291, 320–4, 352
Arup 185–9
Atkins Global 229, 233, 302

Barratt Homes 298
Bayfordbury Estates 308
Bee Bee Developments 325
Big Society 20–2, 104
biodiversity 222, 326, 340, 348
Bishopsgate Goodsyard Regeneration
    167, 266
blogs 46, 49–50, 57–9, 243–4, 250, 254
BME Groups 246, 294–5
BNRG Renewables 218, 349
Bramley Baths 128–9
British Land 256, 318, 326, 347–8
Brooke Smith Planning 265–6, 288–9

Caborn Principles *see* call-in powers
call-in powers 163–4, 175
campaign groups 39–41, 53, 149–51,
    157, 170, 278–90
carbon saving initiatives 326, 334,
    348
Cathelco Group 265–6
CEEQUAL 335
Central England Co-Operative
    288–9
Change of Use Consent 33, 73, 108
charitable donations 327, 347
charrettes 195, 227, 235, 240
Chelsea Barracks Partnership 260
citizen control 3
Cockermouth Neighbourhood
    Development Order 107
collaborative planning *see* participatory
    planning
Committee on Standards in Public Life
    13, 148, 151, 156
communities of interest 38–9
community arts worker 291, 320
Community Asset Transfer
    128–9
community buildings 219, 322, 324–5,
    327, 343–5, 347
community centre *see* community
    buildings
community engagement, definition 3
community funds 345–6
community infrastructure levy 11, 19,
    27, 109
community involvement, definition 3
community liaison officer 327–8
community liaison panel 328–9
community planning *see* participatory
    planning
community relations representatives
    see community liaison officer

community relations website 329–32
community research 42
community rights 121–31
Community Right to Bid 25,
    126–7, 130
Community Right to Build 25, 59,
    121–5, 203
Community Right to Build vanguards
    124
Community Right to Challenge 25,
    125–6, 130
Community Right to Contest 25, 129
community visioning *see* participatory
    planning
Companies Act 347
Complaints Choir 40
Concerto Model 205–7
conservation groups 10, 39, 278
conservation societies *see* conservation
    groups
Considerate Constructors Scheme 316,
    333, 336, 351
construction impacts group *see*
    community liaison panel
construction management plan
    (CMP) 313
consultation: definition 3; stages of 4
consultation fatigue 42, 101, 184,
    292–3, 356
consultation mandate 194, 212–13,
    215, 218, 247, 281
consultation report 196, 274–6, 287–8;
    for NSIPs 180–1, 183–4, 185–6,
    193, 223n1
consultation strategy: aims objectives
    207–8; identity 215; issues analysis
    204–7; messages 208–10; NSIPs
    181–3; principles 217–23; process
    193–7; resourcing a consultation
    214–15; strategic framework 193–4,
    196–7, 216–17; strategic overview
    210–11, 212–13
ConsultOnline 61, 228, 231, 243–4,
    330, 332
content in online consultation 249–50
Convoys Wharf, Lewisham 165–7
corporate social responsibility (CSR)
    41, 338–54
correcting misapprehensions 221–2
Countryside Properties 321
Crossrail 77, 264–5, 270, 315–17, 328
CSR *see* corporate social responsibility
customer experience 349

Data Protection Act (1998) 184
David Lock Associates 227–8, 325, 352
David Wilson Homes 319–20
delegated power 3, 7
demographic data (in consultation
    responses) 231
Department of Communities Local
    Government (DCLG) 87, 109–10,
    115, 175
developer trusts *see* community funds
development consent order 121, 125,
    177, 185
development forum *see* community
    liaison panel
Development Management Procedure
    Order 27
devolution deals 32
digital service standard 51
direct action groups 39–41
disappointing results 290–1
diversity 37, 200, 294–5, 300, 302,
    305, 347
duty to cooperate 28, 84, 86, 91, 94, 100

East Architects 256–7
East Cambridgeshire District Council
    286–7
East London Community land trust
    320–1
education 185, 296, 317–19, 326, 327,
    345, 347
Egan Review: Skills for Sustainable
    Communities 338–43
electricity transmission line, planning
    application for 177
employee engagement 347, 351
employee retention 347, 351
employment initiatives 285, 317–19,
    326, 329
environmental considerations 28, 81,
    88, 158, 199–200, 222, 340
environmental initiatives 319–20, 334,
    347–9
essential land 195
essential living 150, 164, 195, 330
evaluation 230, 270–4, 282–3, 316,
    336–7, 351

Facebook 52–6, 61–4, 144, 228, 244,
    251–3, 308
facilitator 206, 214, 228, 229, 284, 295
feedback: NSIPs 182, 188–9, 216; in
    online consultation 243; tactics

149–50; using feedback 211; within the strategic process *196, 197*, 259; *see also* SCIs
Ferring Parish Council 123–4
Freedom of Information Act (2000) 16, 41
front loading 29–30

Gallagher 342–3
Geddes Institute for Urban Research 257–8
Gershon Report 15
governance 122, 339, 349–50
Greater Manchester Combined Authority 32
Grimsby Docks 170
Growth, Infrastructure Act (2013) 32
Gunning Principles 277
Gunsko Communications 199, 218

harbours, planning applications for 177
hard to reach groups 71, 129, 181, 210, 216, 242, 272, 294–305
Heathrow Airport 2, 278
highways, planning applications for 177
Horizon Nuclear Plant, Anglesey 202, 229, 233, 302
Housing, Planning Act (2016) 28, 33–4, 91, 105, 112
HS2 77–9, 97
human rights 106, 349
hybrid bill 77–9
hyperlocal website 57–61

Information Commissioner's Office 231
Infrastructure Planning Commission (IPC) 19, 177
intelligent community planning *see* participatory planning
internet 38–43, 46–65, 226, 241, 258, 279, 300, 356
internet communication 36–40; in campaigning 53–7; misinformation 49–57; speed of 63, 49, 52, 56; use by central government 50–2; use by local communities 57–61; use by local government 52–3; use by the development industry 61–4, 241–58, 305–7
internet usage (statistics) 46–9, 62
issues analysis 204–7

issues database 283
issues register *see* issues database
Ivy House 127

judicial review 169–71, 175

Kier Group 319, 327, 336, 350–1; Kier Construction 319; Kier Living 327

labour standards 349
Land Use Consultants (LUC) 221, 346
letters of objection (to local authority) 279, 286–7
Lifting the Burden white paper 11
Linden homes 320
local charter 326, 348
Local Development Framework (LDF) 11, 17, 28
local development plan 10, 92–5, 106
local enterprise partnerships (LEPS) 28, 204, 217
Local Government (Access to Information) Act 41
localism 1, 10, 24–8, 34, 44, 53, 59, 100–2, 139, 147, 163; CIL 27; community rights (*see* community rights); Community Right to Bid (*see* Community Right to Bid); Community Right to Build (*see* Community Right to Build); Community Right to Challenge (*see* Community Right to Challenge); Community Right to Contest (*see* Community Right to Contest); duty to cooperate 28; local planning 27–8; Neighbourhood Plan 104–19; predetermination 25, 156; requirement to consult 25
local plan 147–60; consulting on 85; development *86*; duty to co-operate 91–100; in England 82–92; examination 89–92; intervention in plan making 91, 98, 100, 102; issues options 85, 93; key stages 85–8; in National Planning Policy Framework 81, 99; and politics 95–8; preferred options 84, 85, 94, 112; pre-publication 85; in Scotland Northern Ireland Wales 92–5; strategic priorities of 81–2; submission of 89–92; sustainability appraisals 88; time periods *89*
Local Plan Regulations *86*
Local Plans Expert Group 91, 96, 100

local supply chain 325–6
London Borough of Kensington &
    Chelsea 136–8, 144
Love's Farm House 344

McCarthy & Stone 346
managing expectations 291, 182, 221–2
maps (in reporting and analysis) 226–7,
    181, 187, 226, 243, 275
marginalised groups *see* hard to reach
    groups
Marshalls (Holdings) 240
material considerations 19, 69, 150, 160
media relations 307–8
messaging 40, 208–10, 225, 281–2
Middlesbrough Council 140
Milton Park, Oxfordshire 220
Minecraft 257
misapprehensions 205, 215, 221–2, 242,
    261, 278–90
monitoring 215–16, 242, 255, 282–3,
    306, 316–17

National Grid 300–2
National Infrastructure Commission
    177
National Infrastructure Plan 177
nationally significant infrastructure
    projects 177–90; community
    benefits 184–5, 210; consultation
    report 183–4; history 33; process
    178, 180, 216, 282; stakeholders
    180; Statement of Community
    Consultation (SoCC) 180–1
National Planning Policy Framework
    28–32; early engagement 29;
    front-loading 70; local plans 81;
    presumption in favour of sustainable
    development 29, 117, 149; Select
    Committee Inquiry into 31
national policy statement (NPS) 19,
    180–90
negative sentiment 278–90, 314
negotiation 285–6
neighbourhood area 33, 105–7, 113,
    122, 123
neighbourhood development order
    105, 107
neighbourhood forum 105–6, 112–14,
    122, 126, 184, 202, 204
neighbourhood planning 104–19; duty
    to support 107; financial support
    incentives 109–10; judicial review
    119; and national planning policy

117; neighbourhood area 105;
    process 113; referendum 11; take-up
    110–11
Newcastle City Council 53
NewRiver Retail 116, 142, 157
Next Plc 143, 228
NIMBY 279, 278, 309n1
Nolan Principles 13, 148
Northern Ireland 59, 69, 76–7, 92–4,
    105, 168
North West Cambridge 235, 323
NPPF *see* National Planning Policy
    Framework
NSIPs *see* nationally significant
    infrastructure projects

older people 75, 297–8, 300, 302–18
online consultation 241–58;
    accessibility 53, 232, 242, 293, 294,
    298; promoting 254; risks 305–7
online forums 228, 251, 295, 306
online political campaigns 54
open-ended questions 262; *see also*
    qualitative data
Open Public Services white paper
    (2011) 25
opinion polling 143
opposition 38–40, 49, 55, 57, 149, 150,
    201, 278–90

participation statement 92–3
participatory planning 226, 237–41,
    287
PB Planning 280, 298, 319–20
people in planning *see* Skeffington
    Report
permission in principle (PIP) 34
permitted development rights 33,
    73, 132
PEST analysis 198, 208–9, 280
Peter Brett Associates 221, 284
petitions 40, 56, 78, 150, 151, 230, 287
Planning, Compulsory Purchase Act
    (2004) 15, 17, *86*, 91, 274
Planning Act (1947) 9
Planning Act (2008) 19, 27, 177,
    183, 187
Planning Act (Northern Ireland) 76
Planning Act (Wales) 69
planning application: conduct of
    members and officers 151–3;
    declaration of interests 153–6;
    determination 145–6; environmental
    considerations 78, 186, 187, 199,

213, 218, 222, 271; key stages of
133–46; lobbying of members 156–8;
members' role 147–51; officers'
role 147–51; political sensitivity 150;
predetermination 25, 133, 141–5,
147, 153, 156; process 132–46;
role of local authorities 133–5,
140–60, 163–5; statistics 133; tactical
considerations 159–60
planning committee: code of practice
152; register of interests 98, 154
planning inspector *see* Planning
Inspectorate
Planning Inspectorate 89–90, 94,
162–3, 164, 177–8, 184, 188
Planning Policy Guidance notes
(PPGs) 11
Planning Policy Statements (PPSs)
11, 12
Planning Portal 52, 77, 134
Plymouth City Council 143–4, 152–5
political audit 203–4
political research 201–4
Polity 228, 243–4
power generating stations 177
pre-consultation dialogue 196–7, 216,
220, 281–2, 284, 291–2
preferendum 226, *232*
presumption in favour of sustainable
development 29, 117
Prince's Foundation for the Built
Environment, (PFBE) 238–9
project liaison group 229
proscribed local authorities 32
public art *see* arts
public inquiries 162–75; England and
Wales 162–7; intervention by the
London mayor 165–7, 164; judicial
review 169–71; Northern Ireland
168; Scotland 167–8; statistics 168–9
public meetings 122, 127, 151, 237, 261,
278, 329
Public Services (Social Value) Act 314

qualitative tactics/data 259–70, 206,
211, 226–9, 232
quantitative tactics/data 211, 226, 232,
263, 316–17
Queensdale Properties 208, 271
questions 230–1

railways, planning applications for 177
reaching younger people 75, 228, 229,
233, 296

regional assemblies 17, 104
regional development agencies (RDAs)
16–17, 23n8, 27–8
Regional Development Agencies Act
(1998) 16–17
register of interests 154
REG Windpower 199–200
reporting 181–4, 196, 229, 244, 255,
260, 287–90
reputation 2, 286–8, 306–7, 314, 347
research: comment on research in
consultation 42–3; examples 220,
221, 239; in online consultation
54–5, 65, 241; role in consultation
strategy 196–207, 216, 279,
291–2, 294; *see also* PEST analysis;
stakeholder analysis; SWOT analysis
right-to-buy 33
risk in consultation 278–309
Royal Town Planning Institute 2, 152
RWE Innogy ltd 221–2

Sainsbury's 285–6, 289–90
Sayers Common 117–18
SCI *see* Statement of Community
Involvement (SCI)
Scotch Corner Designer Village 243–5,
248, 253
Scottish planning policy 69, 76, 92–4,
167–8
Section 106 11, 19, 27, 321, 342–3, 344
Seven Principles of Public Life *see*
Nolan Principles
Sheringham Shoal 345–6
single issue groups 39–40
site management 327–32
situational analysis 196–200
Skeffington Report 10
social media: advent of 38, 261; analysis
43, 50; benefits of 243, 303, 308; in
campaigning 54, 157, 279; drawbacks
of 246; statistics 46, 62, 244; use of
46, 52–3, 55, 61, 63, 85, 96, 102, 144,
150, 228, 251–5, 308, 329–30
South Cambridgeshire District Council
124–5
special interest groups 39–40, 138, 194,
201, 279, 291
sponsorship 327, 345
stakeholder analysis *197*, 200–4, 208
stakeholder engagement software 204
stakeholder group 136–8, 195, 199,
229, 239, 241; *see also* project liaison
group, community liaison panel

stakeholder mapping 203, 180, 199
stakeholder research 200–4, 221,
    280, 315, 328
statement of community
    consultation 178, 180–1, 216
Statement of Community
    Involvement (SCI) 83, 86, 138;
    definition 223n1; examples 77,
    83, 140; introduction of 17, 76;
    standard content 18, 139; *see also*
    consultation report
StickyWorld 256–7
strategy *see* consultation strategy
structure plans 9, 17
supply chain 325–6, 319, 349, 351
sustainability appraisal 88–9, 101
SWOT analysis 198, 208, 280

tactics 224–36; in community
    relations 327–32; information
    225–6; online or offline 54–7;
    online tactics 241–58, 54–5, 61–5;
    qualitative 226–9; quantitative 226;
    selecting 211, 231–5, 249, 284, *225*
TayPlan 257–8
Tendring District Council 96
Terence O'Rourke Ltd 220
Thame neighbourhood plan 112–13
Thames Tideway Tunnel 185–9
Town and Country Planning Act
    (1968) 9, 10
Town and Country Planning Act
    (1990) 11, 164

Town and Country Planning
    Association 10
training 153, 239, 295, 318, 326, 329,
    334, 340–1, 347–8, 348, 351
Transport for London (TfL) 233, 264,
    267, 283
Turley 110, 115–16
Twitter 38, 52, 53–4, 56, 61–3, 253–4

UK government's consultation
    principles 74
University of Dundee 257
urban development corporations 10

virtual reality 256, 356
VUCITY 256

waste, planning applications for 177
waste saving initiatives 326, 334, 336,
    340, 351
water, planning applications for 177
Web 2.0 49, 53, 54, 241
Wigan metropolitan borough 83, *172*
wind farms, planning applications
    for 177
Woodcock Holdings 117
Woolley 205–7

Yorkshire Forward 227–8
young people 63, 75, 223, 228–9, 233,
    242, 244, 257, 296–7, 300, 303,
    308, 327

Printed in Great Britain
by Amazon